Fundamentals of Human Ethics and Agriculture Extension

The Authors

Dr. D. K. Meena did his master degree in Agriculture Extension from SKRAU, Bikaner and Ph.D. degree awarded in Agricultural Extension Education from National Dairy Research Institute, Karnal. He was awarded UGC fellowship in Ph.D. He started his Professional career as Assistant Professor from Sri Karan Narendra Agriculture University, Jobner, Jaipur. He has published 15 research papers, 25 popular articles, 2 books and 4 book chapters. Dr. Meena got Best Scientist (Teaching) award in 2018.

Dr. Nanchhu Ram Meena is presently working as lecturer and Vice-Principal in the LT. Moolchand Meena Agriculture College, Lalsot Dist. Dausa (Raj.) India. He did his UG and PG from RCA, Udaipur and he has completed Ph.D. from MPUAT Udaipur. Dr. Meena has 3 years of experience in teaching, research and extension activities. He has published many research papers in international and national journals and various popular articles in reputed magazines.

Dr. Nidhi is working as subject matter specialist in extension education at KVK, Maulasar, Nagaur, Agriculture University, Jodhpur and she has completed her UG and PG degree from RCA, MPUAT, Udaipur and also she had completed her Ph. D. from JAU, Junagadh department of Agricultural Extension. She has published 10 research papers in international and national journals. She had published 12 popular articles in reputed magazines. She got Dr. Daya Shankar award from RCA alumina association for securing first position in master degree.

Miss. Sonika Sharma is pursuing Ph.D. and awarded M.sc. degree in Agricultural Extension and Communication, from Govind Vallabh Pant University of Agriculture and Technology, Pantnagar. She did her Under Graduate degree from Maharana Pratap University of Agriculture and Technology, Udaipur. She was awarded with National Talent Scholarship during Master degree. She has published many research papers as well as popular articles.

Fundamentals of Human Ethics and Agriculture Extension

Dr. Devendra Kumar Meena
Dr. Nanchhu Ram Meena
Dr. Nidhi
Miss Sonika Sharma

SCIENTIFIC PUBLISHERS

Published by
SCIENTIFIC PUBLISHERS (INDIA)

5 A, New Pali Road
P.O. Box 91
Jodhpur - 342 001 INDIA

ISBN: 978-93-88172-69-1
978-93-88172-70-7 (E-book)

Visit the Scientific Publishers (India) website at
http://www.scientificpub.com
Printed in India

Preface

This book is prepared for under graduate agricultural sources of 'Fundamental of Agricultural Extension and Human Ethics' as per revised curriculum in Extension Education for B.Sc.Ag. Degree programme. This book will fulfill the needs of ICAR's restructured course curricula and syllabi of extension education and human ethics at the national level. The book is divided into thirty two parts, each dealing with separate aspects. All these parts have been presented in a logical sequence. No books extensively deal with the new course. So, to solve this problem of under graduate students studying under agriculture system, this textbook has prepared. It is expected that the book will meet the requirements of the new system with fullest satisfaction to the students. The 5th Dean's committee recommended the present course curricula and this book is prepared as per the new contents. Earlier human ethics and values was not included as a part of the degree in agriculture. Therefore, informally we learned moral, values and ethics by the parents, friends and teachers etc.

The core areas like recent development program/schemes, concepts, objectives, terminologies in extension, community development programmes, agricultural development programmes, Panchayat raj systems, philosophy and principles of Extension Education; important rural development programmes both are pre and post independent era; human ethics, goal, mission, aims, motivation have been described with suitable illustration simple understandable language for the readers.

An attempt has been made to cover all those relevant aspects, which are very often required for consultation by agricultural Extensionists. The deliberate attempt was made to include newer areas and concept for the benefit of the readers. We acknowledge the pioneers whose work has been cited and to all those who motivated to bring out this publication.

Students, Teachers and professional in extension education and human ethics shall find the book very useful. The book is an essential reading for transforming knowledge in to action.

Authors

Contents

1

Recent Schemes

1.1 STUDENT READY (started in 2015)

The term "READY" refers to "Rural Entrepreneurship and Awareness Development Yojana". Student READY is Skill development initiative to strengthen students with skills, so as to enable them to tackle global challenges, and to improve both their employability as well as ability to set up a venture. Student READY concept signifies this as a finishing school for the undergraduate students.

The students get experience of working on farm in coordination with research stations and KVKs under RAWE Component. The students also stay in villages with farm families, agro-based industries, cooperatives during phases of the RAWE programme to enable them to get real life field experience, understating of the problems and enable them to gain confidence to tackle these problems. Approximately, 25,000 graduates as **Student READY** will be a mandatory requirement for the UG Degree. It is an all inclusive approach to strengthen the entrepreneurship development scenario in the country which is competent, quality conscious, market survey, innovative and has globally competitive entrepreneurs shall be carefully mentored and encouraged. The programme will attract youth towards agriculture and allied sector, and such ventures, when established will help to improve economic conditions in rural areas.

All out efforts are being made to raise profitability of agriculture for making it a really attractive occupation, especially for youth.

1.2 ARYA

Attracting and Retaining Youth in Agriculture: To make agriculture and rural professions intellectually stimulating and economically rewarding to enable to attract and sustain rural youth in agriculture and allied sectors, ARYA programme of ICAR proposed in the July 25th, 2015 during XII Five-Year Plan is to build capacity of rural youth through special programmes and projects including 'Learn while You Earn' programme; develop a comprehensive policy for development of youth in rural areas and recognize the requirements of the new-age farmers and endeavour to fulfill the same. The KVKs (Krishi Vigyan Kendras) will provide training to youth below 35 years of age in different professional skills related to agri-business, help them to prepare their project reports and provide some seeds, money to help them to start their own ventures. A total of 10,000 youth will be covered in the first phage. ARYA project is implemented in 25 states through KVKs, one district from each state.

Basic things about ARYA

- The new scheme ARYA has recently been launched by the Indian Council of Agricultural Research.
- This program is planned to be implemented through Krishi Vigyan Kendra's (KVK-Farm science centres) in 25 states of our country.
- Each KVK would train about 200 to 300 youth in taking up agriculture's allied and supplementary activities, such as poultry farming, dairying, fisheries, goat rearing, mushroom production and other similar activities which keep the rural youth attached to agriculture, either directly or indirectly.
- Finally, the trained young entrepreneurs would be assisted in preparing project reports for seeking bank loans.

Advantages

- Improves Agro-productivity in the country.
- Controls rural migration of youth.
- Income of rural youth becomes sustainable and meaningful.
- Develops service sector in the rural area.
- Business and Entrepreneurial skills of youth get developed.

Conclusion

- Agriculture is the backbone of India.
- ARYA scheme has all the potential to keep the educated youth within the rural areas.
- However, its success depends upon the improvement in the quality life of the rural area.

Objectives of ARYA

(i) To attract and empower the youth in rural areas to take up various agriculture, allied and service sector enterprises for sustainable income and gainful employment in selected districts.

(ii) To enable the Farm Youth to establish network groups to take up resource and capital intensive activities like processing, value addition and marketing.

(iii) To demonstrate functional linkage with different institutions and stakeholders for convergence of opportunities available under various schemes/program for sustainable development of youth.

1.3 FARMER FIRST (started by ICAR, 2015)

Farmer-Farm Innovation Resources in Science and Technology (FIRST): Extension coverage is important for effective technology dissemination. The ICAR is performing this role through nationwide network of Krishi Vigyan Kendra's mandated for frontline demonstration of new technologies. In order to improve the technology dissemination 'Farmer FIRST' is contemplated to enrich farmers-scientists interface for technology development and application with the primary objective to take up technology development based on feedback with the participation of various stakeholders specially farmers. The project focuses on enabling interaction of scientists with farm conditions and problems; exchange of knowledge between farmers and other stakeholders; integration of technology with different agro-systems; development of rural based institutions; and using the platform of the project having commodity institutions as partners to develop commodity specific contents for enabled knowledge sharing.

The initiative will focus on: (i) Enriching Farmers-Scientists interface, (ii) Technology Assemblage, Application and Feedback, (iii) Partnership and Institution Building, and (iv) Content Mobilization.

1.4 MERA GAON MERA GAURAV

Mera Gaon Mera Gaurav: To have effective implementation of 'Lab-to-land programme' in large scale for taking technologies to the villages, the agricultural scientists will visit at least one village every month and interact with the farmers to have solutions pertaining to farming practice. For this purpose, a central scheme named 'Mera Gaon Mera Gaurav' has been put in systematically. Overall, agriculture in India will continue to play its traditional role of producing food, but also generate employment and income opportunities for a sizeable population in the foreseeable future. However, achieving zero hunger goal in future will be a daunting task due to changing demography, growing urbanization, rising incomes, changing food preferences, dwindling land and water resources; and above all, the impact of climate change. It is only through the application and out reach of new science and technology in agriculture that the country would be able to achieve its mission of a food and nutritionally secure India. It is launched by ICAR to provide farmers with required information, knowledge and advisories on regular basis. Under this scheme, groups of scientists will select villages and will remain in touch with that village and provide information to farmers on technical and other related aspects through personal visits or through telecommunication. The scheme enables 20,000 scientists of National Agricultural Research and Education System (NARES) to work directly in villages.

Important question related to topic

- STUDENT READY – **Student Rural Entrepreneurship and Awareness Development Yojana**
- 25000 graduates as **Student READY** will be a mandatory requirement for the UG Degree
- **ARYA - Attracting and Retaining Youth in Agriculture**
- ARYA programme of ICAR proposed in the XII Five-Year Plan
- The KVKs will provide training to youth below 35 years of age in different professional skills related to agri-business
- ARYA project is total of 10,000 youth will be covered in the first phage
- ARYA project is implemented in 25 States through KVKs, one district from each State.
- ARYA project implemented KVK, Pali in Rajasthan
- ARYA has recently been launched by ICRA
- FARMER FIRST – **Farmer-Farm Innovation Resources in Science and Technology**
- **Mera Gaon Mera Gaurav** – To have effective implementation of 'Lab-to-land' programme.

■■■■

2

Education

2.1 MEANING AND DEFINITION

The word *"Educate"* comes from the Latin *'Educere'*, meaning "To lead out", "To bring out". Education is the harmonious development and having four dimensions of life *i.e.,* Physical, Mental, Moral and Social.

Definitions of Education: By Webster "Education is the production of desirable changes in the behaviour of the people in term of knowledge (things known), attitudes (things felt) and skills (things done), in all of them or in one or more of them".

2.2 FUNCTIONS/ ELEMENTS OF BEHAVIOUR

1. **Change in knowledge:** Means change in ***"What people know"***. For example, a farmer who does not know about rodent management techniques, may know it through participation in extension programme viz., modern rat traps, use of celphos tablets, fumigation techniques, ITK's and dosages, time, method and place of application of chemicals etc. Similarly, Extension Agent who does not know about Information Technology (IT) comes to know of it after attending a training programme.

2. **Change in skills:** Means change in the ***"Technique of doing things"***. Example, the farmer learnt the technique of Paneer making which he did not know earlier. Calculation

of seed rates, fertilizer dosages and pesticide dosages etc., Skills of women farmers on mushroom cultivation can be improved by demonstrations.

3. **Change in attitude:** Means ***"Change in the feeling towards any object, person and thing".*** For Example, farmer developed a favourable attitude towards the paneer making. Farmers attitude is killing rat is a sin because it is the *Asana* of Lord Ganesh, this attitude has to be changed by explaining them the extent of losses caused by rodents to them and to their neighbours, if left uncontrolled.

2.3. TYPES OF EDUCATION

Coombs and Ahmed (1974) drew three (FIN) types of education.

Formal education: It is highly institutionalized, uniform, full time, chronologically graded, hierarchically structured education system which starts from primary school to university education. **School to college education.**

In-Formal education: It is a lifelong educational process by which every individual acquires and accumulates knowledge, skills, attitudes and insights, from daily experiences and exposure to environment at home, at work, at play etc. *e.g.* **Developing habits based on moving with friends at play**

Non-Formal education: It is also organised and systematic educational activity outside the framework of formal education system to provide selected types of learning to particular sub groups of population, adults, as well as children according to their need. *e.g.*, **Extension Education.**

Difference between Formal Education and Non-Formal Education

Formal Education	Extension Education (Non-Formal education)
The teacher starts with theory and works up to practical	The teacher (extension worker) starts with practical and may take up theory later on.
Students study subjects	Farmers study problems

Formal Education	Extension Education (Non-Formal education)
The curriculum is fixed	It has no fixed curriculum or course of study and the farmers help to course to formulate the curriculum
Authority rests with the teacher	Authority rests with the farmers
Class attendance is compulsory	Participation is voluntary
Teacher instructs the students	Teacher teaches and also learns from the farmers
Teaching is only through instructors	Teaching is also through local leaders
Teaching is mainly vertical	Teaching is mainly horizontal
Homogeneous audience	Heterogeneous audience
It has all pre-planned and pre-decided programmes	It has freedom to develop programmes locally and they are based on the needs and expressed desire of people
It is more theoretical	It is more practical and intended for immediate application in the solution of problems
It is rigid	It is flexible

Important points from the chapter

- **Education word rooted from –** 'Educere' word of Latin language
- **Production of desirable change in people behaviour –** Education
- **Function/function of behaviour –** Knowledge, Skills and Attitude
- **Behaviour components means –** Know, Done and Felt
- **Types of Education –** Formal, In-formal and Non-formal
- **Formal education –** Fixed, rigid, chronologically, hierarchically
- **In-formal education is –** Life long process

- **Non-formal –** Outside the frame work of formal education
- **Example of Non-formal education –** Extension education
- **Formal education –** Rigid, Homogenous, Vertical Fixed curriculum
- **Extension Education –** Horizontal, Flexible, Horizontal
- Differentiate between Formal and Non-Formal Education
- Define the Formal, In-Formal and Non-formal education.

■■■■

3

Extension Education

3.1 MEANING AND DEFINITION

The word extension is derived from the ***Latin*** roots ***'Ex'*** – meaning ***'Out'*** and ***'Tensio'*** meaning ***'Stretching'***. Stretching out is the meaning of extension. The word 'Extension' came to be used originally in USA during 1914 which means *"a branch of a university for students who cannot attend the university proper"*. In other words, the word extension signifies an out of school system of education.

Education is an integral part of extension. The basic concept of extension is that it is education. Extension means that type of education which is stretched out to the people in rural areas, beyond the limits of the educational institutions to which the formal type of education is normally confined.

First agricultural society was established in Scotland in 1728 for the spread of academic knowledge of agriculture through lectures and publication. The first practical steps were taken in 1867–68 when James Stuart, fellow of Trinity College, Cambridge, England gave lectures to women's associations and working men's clubs. Stuart is called as Father of University Extension. He approached the authorities of university of Cambridge to organize extension lectures. The system was accepted by the Cambridge as well as Oxford University in 1873. Dr. J.P. Leagan's of USA is known as the father of extension education, whereas Dr. K.N. Singh is known as father of Extension education in India.

Land Grand College Act or Morril Act was passed in 1962 to start educational opportunities for working men in subject of practical interest. Thus, Land Grand College was broad enough to take education beyond the class room for farmers. After passing of Hatch Act in 1887, one experimental station was established with each Land Grand College.

Definitions of Extension Education

Engsminger, 1957 – Extension is education and its purpose is to change attitude and practices of the people with whom the work is done.

The National Commission on Agriculture, 1976 – Refers to extension as an out of school education and services for the members of the farm family and others directly or indirectly engaged in farm production, to enable them to adopt improved practices in production, management, conservation and marketing.

Van den Ban and Hawkins, 1988 – Extension involves the conscious use of communication of information to help people from sound opinions and make good decisions.

Leagan's, 1961 – Extension education is an applied science consisting of content derived from research, accumulated field experiences and relevant principles drawn from the behavioural sciences synthesized with useful technology in to a body of philosophy, principles, content and methods focused on the problems of out of school education for adults and youth.

- An out-of-school system of education
- An educational process
- To teach a person how to think, not what to think
- A continuous process
- An applied science
- An applied behavioural science
- A science which deal with various strategies of changes in the behavioural pattern of human beings
- A service or system which assists farm people

Extension service – Refers to a program for agricultural development and rural welfare which employees the extension process as a means of program implementation.

Extension process – The working with rural people through out of school education along those lines of their current interest and need which are closely related to gaining a livelihood and improving the physical level of living of rural families as well as fostering rural community welfare.

Agricultural extension – It is known as the application of scientific research and new knowledge to agricultural practices through farmer education.

3.2 SCOPE OF EXTENSION EDUCATION

The scope of extension education includes all the activities directed towards the development of the rural people.

The scope of extension education according to Kelsey and Hearne are as below:

1. Efficiency in agricultural production, marketing, distribution and utilization
2. Conservation, development and use of natural resources
4. Management on the farm as well as home
5. Family living
6. Development of Youth, Leadership, Community and Rural development
7. Improving public affairs for all round development

3.3 PROCESS OF EXTENSION EDUCATION

According Dr. J. Paul Leagan's, Extension Education takes place as a process. The Education starts with study of present situation and identification of problems. Based on the problems identified, solutions are found out and objectives are formulated. In the third phase, plan of work is taught on how to attain the objectives and then the entire process is evaluated and tested whether formulated objectives are attained or not and in the final stage reconsideration about the unattained objectives and unaddressed problems is done.

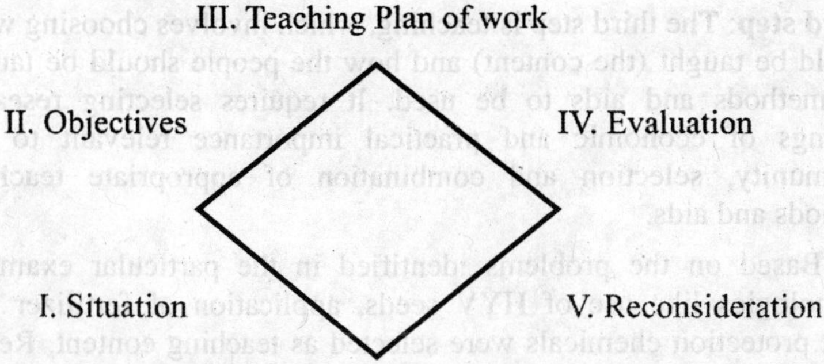

Extension Education Process

First step: The first step consists of collection of facts and analysis of the situation. Facts about the people and their enterprises; the economic, social, cultural, physical and technological environment in which they live and work. These may be obtained by appropriate survey and establishing rapport with the people.

The responses obtained are to be analyzed with the local people to identify the problems and resources available in the community. For example, after a survey in a community and analysis of the data, the problem was identified as low income of the farm family from their crop production enterprise.

Second step: The next step is deciding on realistic objectives which may be accomplished by the community. A limited number of objectives should be selected by involving the local people. The objectives should be specific and clearly stated and on completion should bring satisfaction to the community. Objectives should state the behavioural changes in people as well as economic and social outcomes desired. In the example, the problem was identified as low income from the crop production enterprise. A deeper probe into the date revealed that low income was due to low yield of crops, which was attributed to the use of local seeds with low yield potential, application of little fertilizer and lack of protection measures. By taking into consideration the capacity and competency of the people in the community and the availability of resources, the objective was set up to increase the crop yield by 20 per cent within a certain period of time. It was estimated that the increased yield shall bring increased income, which shall enhance the family welfare.

Third step: The third step is teaching, which involves choosing what should be taught (the content) and how the people should be taught the methods and aids to be used. It requires selecting research findings of economic and practical importance relevant to the community, selection and combination of appropriate teaching methods and aids.

Based on the problems identified in the particular example, technologies like use of HYV seeds, application of fertilizer and plant protection chemicals were selected as teaching content. Result demonstration, method demonstration, farmers' training and farm publications were chosen as teaching methods, tape recorder and slides were selected as teaching aids.

Fourth step: The fourth step is evaluating the teaching *i.e.*, determining the extent to which the objectives have been reached. To evaluate the results of an educational programme objectively, it is desirable to conduct a re-survey. The evidence of changed behaviour should be collected, which shall not only provide a measure of success, but shall also indicate the deficiencies, if any.

In the example, the re-survey after the fixed period of time, indicated that the crop yield had increased by 10 per cent. It, therefore, indicated that there was a gap of 10 per cent in crop yield in comparison to the target (objective) of 20 per cent fixed earlier. The re-survey also indicated that there had been two important deficiencies in carrying out the extension educational program, such as, there was lack of proper water management and the farmers could not apply the fertilizer and plant protection chemicals as per recommendation due to lack of funds.

Fifth step: The fifth step is re-consideration of the entire extension educational programme on the light of the results of evaluation. The problems identified in the process of evaluation may become the starting point for the next phase of the extension educational programme, unless new problems have developed or new situations have arisen.

After re-consideration of the results of evaluation with the people, the following teaching objectives were again set up. For example, they were, training the farmers on proper water management practices and putting up demonstrations on water management.

The people were also advised to contact the banks for obtaining production credit in time to purchase critical inputs.

Thus, the continuous process of extension education shall go on, resulting in progress of the people from a less desirable to a more desirable situation.

3.4 OBJECTIVES OF EXTENSION EDUCATION

Objectives: They are expressions of the ends towards which our efforts are directed.

An objective means ***"Directions of movement"***. Main objective of extension is to work with people as individuals, as families in the home and with their environment.

Types of objectives

1. Fundamental objective: The fundamental objective of extension is the development of the people or the ***"Destination man"***. In other words, it is to develop the rural people economically, socially and culturally by means of education *e.g.,* to increase socio-economic status and standard of living of Indian farming Community.

2. General (Function) objective: The general objectives of the extension are:

1. To assist people to discover and analyze their problems, their felt and unfelt needs.
2. To develop leadership among people and help them in Organising groups to solve their problems.
3. To disseminate information based on research and/or practical experience, in such a manner that the people would accept it and put it into actual practice.
4. To keep the research workers informed of the peoples' problems from time to time, so that they may offer solutions based on necessary research. For example to increase the production and productivity of food grains in Rajasthan.

3. Working objective: It is the one which focuses on specific activity of a specific group in a selected geographic area *e.g.,* to improve the educational status of BA first year students in a University X.

3.5 PRINCIPLES OF EXTENSION EDUCATION

"Principles are generalized guidelines which form the basis for decision and action in consistent way". Following principles are below here :

1. **Principle of cultural difference** – Culture simply means social heritage. There is cultural difference between the extension agents and the farmers. Differences exist between groups of farmers also. The differences may be in their habits, customs, values, attitudes and way of life. Extension work, to be successful must be carried out in harmony with cultural pattern of the people.
2. **Grass root principle** – Extension programmes should start with local groups, local situation and local problems. It must fit to the local conditions. Extension work should start with where people are and what they have. Change should start from the existing situation, *e.g.,* Extension worker should train illiterate farmers initially on marketing aspects rather than training on complicated topic like WTO or GATT.
3. **Principle of interests and needs** – To be effective, extension work must begin with the interest and needs of the people. Many times the interests of the rural people are not the interests of the extension worker. Even though he sees the needs of the people better than they do themselves, he must begin with the interests and needs as they (the people) see them. Only in this way extension agency can mould the needs and interests of the people into realistic requirements. Needs that can satisfy the individuals, groups, community and national interests, needs that can be fulfilled with the available resources should be fulfilled first. *e.g.,* Extension work is successful if it is according to people's needs, demonstration on Paddy cultivation in low lying areas. Extension work fails if it is not according to people's needs, demonstration on Castor cultivation in low lying areas.
4. **Principle of indigenous knowledge** – People every where have indigenous knowledge systems which they have

developed through generations of work experience and problem solving in their own specific situations. The indigenous knowledge systems encompass all aspects of life and people consider it essential for their survival. Instead of ignoring indigenous knowledge system as outdated, the extension agents should try to understand them and their ramifications in the life of the people, before preceding the recommend something new to them.

5. **Principle of learning by doing** – Learning by doing involves use of maximum number of senses; hence it is very effective in changing behaviour, *e.g.,* Demonstration on working of Conon weedier in SRI cultivation is very effective than lecture method. In extension work, farmers should be encouraged to learn new things by doing and by direct participation.
6. **Principle of participation** – The participation of the people is of fundamental importance for the success of any educational endeavour. People must share in the development of a programme and must feel that it is their own programme, *e.g.*, Success of dairy mela at national dairy research institute (NDRI) every year is due to people's participation.
7. **Principle of leadership** – Identifying different types of leaders and working through them essentially in extension. Local leaders are the custodians of local thought and action. The involvement of local leaders and legitimizations by them are essential for the success of a programme. Leadership traits are to be developed in the people so that they of their own shall seek change from less desirable to more desirable situation. The leaders may be trained and developed to act as carriers of change in the villages, *e.g.*, Farmers gets easily convinced about latest technology if it is adopted by a local leader than taught by an extension worker.
8. **Family principle** – Family is the primary unit of society. The target for extension work should, therefore, be the family. That is developing the family as a whole econom-

ically and socially. Not only the farmers, the farm women, youth are also to be involved in extension programmes.

9. **Principle of adaptability –** Extension work and extension teaching methods must be flexible and adapted to suit the local condition. This is necessary because the people's situation, resources and constraints varies from place to place and time to time. *e.g.*, use of mass media like TV, computer and internet to demo an agricultural practice cannot be performed in an interior village where electricity is uncertain. Instead posters, charts, live specimens could be entertained.
10. **Principle of satisfaction –** The end product of extension work should produce satisfying results for the people. Satisfying results reinforce learning and motivate people to seek further improvement, *e.g.*, if a farmer is satisfied by seeing baby corn cultivation in an exposure visit he tries to adopt it.
11. **Principle of evaluation –** Evaluation prevents stagnation. There should be a continuous built in method of finding out the extent to which results obtained are in agreement with the objectives fixed earlier. Evaluation should indicate the gaps and steps to be taken for further improvement.

3.6 PHILOSOPHY OF EXTENSION

According to Kelsey and Hearne (1967) *the basic philosophy of extension education is to teach people how to think, not what to think.* Extension's specific job is furnishing the inspiration, supplying specific advice the technical help, and counseling to see that the people as individuals, families, groups and communities work together as a unit in "Blueprinting" their own problems, charting their own courses, and that they launch forth to achieve their objectives. Sound extension philosophy is always looking ahead.

3.7 PRINCIPLES UNDERLYING THE PHILOSOPHY OF EXTENSION

1. Extension is an organisation to plan, execute and evaluate programmes with the people, and not for the people.

2. Extension is an organisation set up to teach people and motivate them to action, not to dictate what people should do.
3. Extension should help people to help themselves.
4. Extension should be based on felt needs and enlightened desires of the people.
5. Extension should reach the people where they are.
6. Extension aims and objectives should not be rigid, but it should be flexible (Time, date etc.)
7. Extension should change the people and not the subject matter.
8. Extension should work in harmony with the culture of the people.
9. Democratic procedures must be adopted in the formulation and execution of the programmes (group ideas only)

Points to be remembered

- **Extension from derived** - Latin word
- **'Ex'** - Out, **'Tensio'** - Stretching
- **Extension came to be used originally** - In USA during 1914
- **First practical step taken in and by** - 1867–1868, James Stuart
- **Father of University Extension** - James Stuart
- **Father of Extension Education** - J.P. Leagan's
- **Father of Extension Education in India** - K.N. Singh
- **When was passed Land Grand College or Morril Act** - 1962
- **Hatch act was came in** - 1887
- **Extension** - Educational process, Continuous process, Applied Science, An out of school system
- **Direction of movement** - Objective

- **Objective** - Are expression of the ends towards which our efforts are directed
- **Types of Objective** - Three (Fundamental, General and Working)
- **Culture is** - Social heritage
- **Primary unity of society** - Family
- **Evaluation prevents** - Stagnation
- **Basic philosophy of extension** - is to teach people, how to think, not what to think
- Describe the extension educational process with example
- Define the - Extension Service, Extension Process and Agricultural Extension.

■■■■

4

Extension Programme Planning

4.1. MEANING AND DEFINITION OF PROGRAMME PLANNING

Programme development is an on-going dynamic process that extension professionals intuitively follow as they plan, implement and evaluate their educational programmes. It is applied on a small scale as an individual workshop is being developed, on a larger scale as a series of programmes are being developed around a single priority and on an even larger scale at district level and state level wide team sets priorities and defines a plan of action for a five-year period.

Extension programme

According to Kelsey and Hearne (1949), an "extension programme" is a statement of situation, objectives, problems and solutions.

So, we can define an **extension programme** as a written statement of situation, objectives, problems and solutions which has been prepared on the basis of an adequate and systematic planning effort and which forms the basis of extension teaching activities in a specific area, for a given period.

Programme is a written statement containing a more pertinent factual data used in decision-making, the problems agreed upon with priority assignment and the possible solutions to the problems.

Plan or Plan of work is an outline of activities so arranged as to enable efficient execution of the entire programme. It answers the questions of what, why, how, when, where and by whom the work is to be done.

Project is a single item of the annual plant containing the method of solution of a single selected problem.

Calendar of work is a plan of work arranged chronologically, according to the time when step of work is to be done. It is a time schedule of work.

Aim is a broad objective. It is a generalized statement of direction and may have several objectives. It is also said to be an end in view to give direction to the creative process.

Objective is a direction of movement. A well stated objective is always measurable. It is also said to be a goal of growth.

Goal is a distance in any given direction, proposed to be covered in a given time.

Planning is designing a course of action to achieve ends.

Programme planning

Programme planning is the process whereby the people in the country through their leaders plan their extension programme, country and state professional extension staff member assist in this process (Lawrence, 1962).

Programme planning is the process of making decisions about the direction and intensity of extension-education efforts of extension-service to bring about social, economic and technological changes.

4.2. STEPS/PROCESS OF PROGRAMME PLANNING

Program development involves: analyzing the situation or context; setting objectives/priorities and designing an action plan; implementing the plan; evaluating inputs, outputs, outcomes, and reconsider the weak and strong points of the programme.

The program development process is on-going and continuous. Each educational initiative, workshop or event we carry out modifies the initial situation. As a consequence, any plan of action continues to evolve and change as the situation or context changes.

Collection of facts

Sound plans are based on availability of relevant and reliable facts. This includes facts about the village people, physical conditions, existing farm and home practices, trends and outlook. Besides, other facts about customs, traditions, rural institutions, peoples' organisations operating in the area, etc. should be collected. The tools and techniques for collecting data include systematic observations, a questionnaire, interviews and surveys, existing governmental records, census reports, reports of the Planning Commission, Central Bureau of Statistics, and the past experiences of people.

Analysis of the situation

After collecting facts, they are analysed and interpreted to find out the problems and needs of the people.

Identification of problems

As a result of the analysis of facts, the important gaps between 'what is' and 'what ought to be' are identified and the problems leading to such a situation are located. These gaps represent the peoples needs.

Determination of objectives

Once the needs and problems of the people have been identified, they are stated in terms of objectives and goals. The objectives represent a forecast of the changes in the behaviour of the people and the situation to be brought about. The objectives may be long-term as well as short-term, and must be stated clearly.

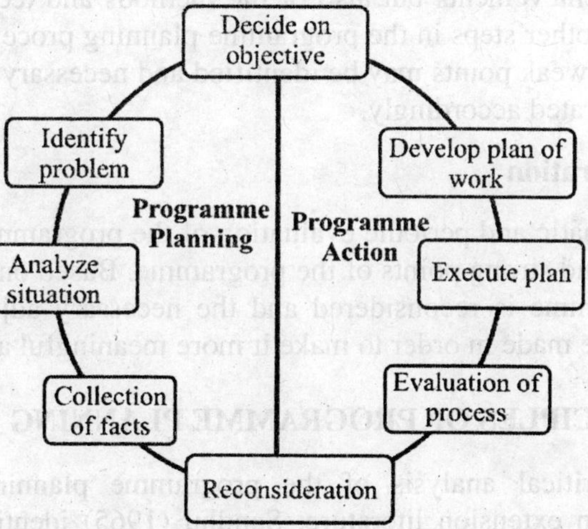

Steps of programme planning

Developing the plan of work

In order to achieve the stated objectives and goals, the means and methods attaining each objective are selected; and the action plan, *i.e.* the calendar of activities is developed. It includes the technical content, who should do what, and the time-limit within the work will be completed. The plan of work may be seasonal, short-term, annual or long-term.

Execution of the plan of work

Once the action plan has been developed, arrangement for supplying the necessary inputs, credits, teaching aids, extension literature etc. has to be made and the specific action has to be initiated. The execution of the plan of work is to be done through extension methods for stimulating individuals and groups to think, act and participate effectively. People should be involved at every step to ensure the success of the programme.

Evaluation

It is done to measure the degree of success of the programme in terms of the objectives and goals set forth. This is basically done to determine the changes in the behaviour of the people as a result of the extension programme. The evaluation is done not only of the physical achievements but also of the methods and techniques used and of the other steps in the programme planning process, so that the strong and weak points may be identified and necessary changes may be incorporated accordingly.

Reconsideration

The systematic and periodic evaluation of the programme will reveal the weak and strong points of the programme. Based on these points, the programme is reconsidered and the necessary adjustments and changes are made in order to make it more meaningful and sound.

4.3. PRINCIPLES OF PROGRAMME PLANNING

After a critical analysis of the programme planning principles available in extension literature, Sandhu (1965) identified a set of principles that may be applicable in developing countries.

Extension programme planning is based on analysis of the facts in a situation

It is important to take into account the conditions that exist at a particular time. This implies that factors such as land, crops, economic trends, social structure, economic status of the people, their habits, traditions and culture, in fact, everything about the area in which the job is to be done and its people, may be considered while planning an extension programme for an area.

Extension programme planning selects problems based on people's interests and needs

Sound programme building selects problems based on people's needs. It is necessary to select these problems which are most urgent and of widest concern. Choice of problems must be from among those highlighted by an analysis of the facts regarding what are felt as unfelt needs. It is common knowledge that people join together because of mutual interests and needs.

Extension programme planning determines objectives and solutions

In order to hold interest, we must set working objectives and offer solutions which are within reach and which will give satisfaction on achievement. This is related to motivation for action. People must see how they or their communities are going to benefit from the proposed solutions. Very often the simplicity or dramatic effect of the practice recommended is the most potent factor in its wide adoption. Further, if there is to be progress and not more evolution in the development of man, the objectives must be periodically revised in view of the progress made. In other words, as changes occur, objectives need to be re-determined to allow for even further progress to be realized.

Extension programme planning has permanence with flexibility

Any good programme must be forward looking and permanent. Permanence means anticipating years of related and well organised effort. Along with this lower process, which both follow and makes a long-term trend, experience has shown that particular items will need to be changed to meet unforeseen contingencies or emergencies. Without flexibility, the programme may not, in fact, meet the needs of the people. A programme should be prepared well in advance of its execution, but not too far ahead of time.

Extension programme planning has balance with emphasis

A good programme should cover the majority of people's important interests. It must be comprehensive enough to embrace all groups, races at all levels and community, block, state, national and international problems. It is futile to deal with only one phase of life in a community as an end in itself. At the same time, a few of the most important or timely problems should be chosen for emphasis. To avoid scattered effort, something must stand out. Decisions must be made as to which of the needs are most urgent.

Extension programme planning has a definite plan of work

No matter how well a programme is through, it is of no use unless carried out. This implies good organisation and careful planning for action. A plan of work is an outline of procedure so arranged as to enable efficient execution of the entire programme. It is the answer to what, where, when and how the job will be done.

Extension programme planning is an educational process

The people who do the planning may participate in local surveys and neighbourhood observations. This provides an opportunity for them to learn more about their own community and area, and increases their interest. The extension worker has the responsibilities of providing local leaders with the knowledge, skills and attitudes they must have if they are to help in educationally serving the people. Essentially, learning takes place through the experiences the learner has and the responses he makes to the stimuli of his environment. The experience gained in finding facts, analyzing situations, recognizing problems, stating objectives and thinking of possible solutions and alternatives should make for a better and more effective learning environment. The extension personnel should remember this fact and provide opportunities for the effective participation of local people in programme planning.

Extension programme planning is a continuous process

Since programme planning is viewed as an educational process and since education is seen as a continuous process, therefore it logically holds that extension programme planning is a continuous process. There is no question of exhausting new knowledge, either in the subject matter with which we deal or in the methods of teaching. With the constant flux of agricultural technology, extension education is faced with an increasingly more difficult job as it tries to serve the needs and interests of the people. He set forth five steps within might be useful in making necessary adjustments:

- Keep choice to the people
- Be flexible and ready to grasp with firmness new problems as they arise.
- Work with people in seeking practical solutions to their problems.
- Keep abreast of technological and social change.
- Close the gap between research discovery and practical application.

It is obvious that tomorrow's problem will not be the same as today's. So extension must make periodic adjustments in its plans to meet the changing problems. Extension must also be alert to the change that is going on in Science and Technology. With new technology, solutions to problems change. It is therefore necessary to view extension programme planning as a continuous process though its recurrence is cyclic.

Extension programme planning is a co-coordinating process

Extension programme planning finds the most important problems and seeks agreement on definite objectives. It coordinates the efforts of all interested leaders, groups and agencies and considers the use of resources. It obtains the interest and co-operation of many people by showing them why things need to be done. This is important in working with people. Within the extension organisation, the block staff may work together on an integrated programme, each member devoting part of his energy to appropriate phases.

Extension programme planning involves local people and their institutions

Involvement of local people and their institutions is very essential for the success of any programme for their development. People become interested and give better support to the programme when they are involved in the planning process.

Extension programme planning provides for evaluation of results

Since extension programme planning involves decision-making procedures, so evaluation is important in order to make intelligent decisions aimed at achieving the stated objectives. Matthews (1962) pointed out that extension programme planning and evaluation go together. Kelsey and Hearne (1949) have said that all other principles of programme building are related to evaluation. Effective evaluation will, of course, depend on clear objectives, knowing which people we are trying to teach and having records of the results in terms that reflect changes in their action. Starting a programme with the intention of engaging in a careful evaluation at the close of a specific period has a salutary effect on all the intermediate processes. However, provision has to be made both for concurrent and ex-post facto types of evaluation.

Important point related to topic

- The statement of situation, objectives, problems and solutions – **Extension programme**
- plan of work arranged chronologically – **Calendar of work**
- Direction of movement – **Objective**
- Goal – **A distance in any given direction, proposed to be covered in a given time**
- Steps of programme planning – **Eight**
- One to four steps – **Programme implementation**
- Five to Eight steps – **Programme Determination**
- Explain the process of programme planning
- Enlist the principles of programme planning

■■■■

5

Monitoring and Evaluation

5.1. CONCEPT AND DEFINITION

Extension programmes are mostly funded with public money and are planned and implemented by an organisation, which in most cases is a department of a government. In order to justify the appropriation of public funds and continuing support from the people, it is necessary that their management as well as impact be properly and adequately evaluated from time to time. How, to evaluate management, achievements and failures of these programmes has been a challenge to extension workers right from the time when planned extension programmes were introduced. However, 'it was when Tyler's (1950) philosophy of educational evaluation became a part of extension education that educational evaluation became a part of extension education that the pattern of extension educational evaluation took a more usable, understandable form' (Sabrosky, 1966).

The word 'evaluation' has its origin in the Latin word "Valere" meaning to be strong or valiant. Its dictionary meanings are the determination of the value, the strength or worth of something, an appraisal and estimates of the force of or making a judgment of something.

Evaluation as applied to the field of extension education may be defined as "a process of systematic appraisal by which we determine the value, worth or meaning of an activity or an enterprise". It is a method for determining how far an activity has progressed and how much further it should be carried to accomplish objectives. Thus to an extension worker evaluation means determining the results of his extension programmes in order to know the extent to which objectives have been achieved and why and what changes would be needed in case the programme is planned again, or in its implementation.

Tyler (1950) developed two basic notions regarding educational evaluation, which equally apply to extension evaluation. These notions are that the process of evaluation (i) is essentially a process of determining behaviour of the people covered under the programme and (ii) the process of determining the degree to which these behavioural changes are actually taking place. Thus extension evaluation may be said to be a process for determining behavioural changes of people resulting from extension programmes. Once evaluation became an integral part of the extension education process, extension managers started applying this process to evaluate programme planning, management and implementation aspects of extension programme.

Definitions of evaluation

More specific definitions of evaluation are given by persons involved in rural development programmes. While most of these definitions refer specifically to the assessment of the results of programmes of extension education, they can also be applied to the training aspect of such programmes. Some definitions of evaluation are:

1. It is a process, which enables the administrator to describe the effects of his programme and thereby make progressive adjustments in order to reach his goal more effectively (Jahoda and Barnit, 1955).
2. Programme evaluation is the determination of the extent to which the desired objectives have been attained or the amount of movement that has been made in the desired direction (Boyle and Johns, 1970).
3. Programme evaluation is the process of judging the worth or value of a programme. The judgment is formed by comparing the programme should be (Steele, 1970).
4. Evaluation is the process of delineating, obtaining and providing useful information for judging decision alternative (Stuffle Beam, 1971).

5.2 MONITORING OF EXTENSION PROGRAMMES

Nature of evaluation

1. Evaluation is not measurement: Evaluation is an integral part of extension education. All aspects of extension work need evaluation. Evaluation does not mean mere measuring of achievements, which is usually done after the programme is executed. Extension being an educational process, it is necessary to evaluate management of the programme and methods used, achievements accomplished in line with the objectives and also to determine the reasons for success or failure.

2. Evaluation is not exactly scientific research: When we think of evaluation as a process of collecting information as a basis for making decisions, forming judgments and drawing conclusions, we realize it has much in common with scientific research. But there is a great difference between our casual every day evaluation and scientific research. However, the difference is a matter of degree rather than kind. Casual every day evaluation can be placed at one end of the scale and scientific research at the other end. There are five locations on the scale with no sharp lines of distinction, *i.e.*, casual every day evaluation, self-checking evaluation, do-it yourself evaluation, extension evaluation studies and scientific research.

1	2	3	4	5
Casual everyday evaluation	↕ Self checking evaluation	Do ↕ it yourself Evaluation	Extension evaluation studies	Scientific research

Informal ⟶ Formal

5.3 EVALUATION OF EXTENSION PROGRAMMES

Types of evaluation

1. Self-evaluation: This is to be carried out by every worker as a matter of routine. This requires the self-critical attitude which is so essential for extension work. By this self-critical attitude, the chances of an extension worker growing and continuously improving his professional competency become greater.

2. Internal evaluation: Evaluation carried to, by the agency responsible for the planning and implementation of the programme. Some of the other methods for internal evaluation are: systematic use of diaries and reports of workers, planned visits of staff members to work spot, use of special questionnaires and Performa for observation and inquiry etc.

3. External evaluation: Evaluation conducted by a person or a committee outside the area of operation. One of the strong features of the Indian Community Development Programme is that simultaneous with its start an independent agency, namely the Programme Evaluation Organisation, was established.

Evaluation can also be classified into (i) con-current and (ii) ex-post facto evaluation

Evaluate programme planning

As a result of experience, theory, research and experimentation, much information has been accumulated about how an extension programme should be planned. Progress in science and technology and the broadening of extension's clientele with the accompanying great variation in needs and interests have made the scientific planning of extension education programmes more important than ever before. There is considerable agreement on certain criteria which, if followed, make for successful extension programme planning at different levels. These criteria represent the ideal with which to compare our practices and procedures or programme planning. Some of the steps needed to evaluate or programming function in view of these criteria include:

i. Identify the evidence needed to form a judgment about each criterion.

ii. Specify the methods that will be used to obtain the evidence, such as personal observation, personal interview or through a systematic survey.

iii. On the basis of the evidence gathered, judge whether or not each criterion is being adequately satisfied in the programme planning activities.

Extension evaluation process

There are several models of evaluation available in the literature. However, a very simplified version of most of these models may be quite workable for evaluating extension programmes since, as Bhatnagar (1987) has pointed out, any extension evaluation process has to be based on certain assumptions. For example, if some inputs are provided in the form of a programme, specific outputs can be expected and if these outputs happen, then the purpose of the programme can be achieved; if the purpose is achieved, then the development goal is realized. This means that evaluation has to be so designed that the quality types and adequacy of the input measures, outputs and their impact in achieving the programme objectives have to be evaluated systematically.

Steps involved in an extension programme evaluative process may be as follows

(i) Formulate evaluation objectives

Specific objectives to be achieved through the evaluative process must be clearly and adequately identified and started. All further efforts should be knit around these objectives.

(ii) Classify programme objectives

It is assumed that each extension programme, when formulated and implemented, will have specific well-defined objectives. Since evaluation is basically a process of determining the extent to which various extension teaching activities were organised and managed and the extent to which they contributed to achieving the goals, programme objectives must be clearly understood and if necessary, further broke down into measurable terms. This is a crucial step as all further efforts will be directed towards collecting evidence related to these objectives.

(iii) Identify indicators

To identify indicators or the kind of evidence necessary to evaluate achievement in relation to specified programme objectives, it is necessary that specific beneficiaries of the programme be identified, the kind of behavioural changes expected in them be clearly stated, and the kind of learning experiences expected to be provided to them spelled out, together with the level of management to be achieved for provided those learning experiences are specified. Once this is done, identification of specific indicators to measures the achievements will not be difficult.

(iv) Decide the kind of information needed

Once the indicators for evaluating the management and performance of a programme have been indicated, specific information to be collected may be worked out. Since there is usually more information than an extension worker can collect, he has to be very discriminating about the kind and amount of information that should be collected. Timing for collection of information may also need to be specified.

(v) Sampling

The purpose of sampling is to take a relatively small number of units from a population in such a way that the evidence collected from them becomes representative evidence of the entire population. Although there are several sampling methods, perhaps stratified sampling procedures may be most suitable for extension evaluation studies, as they allow inclusion of all interested groups and ensure enough heterogeneity in the sample.

(vi) Decide the design of evaluation

An ideal design of evaluation may be an experimental one. This would allow separating the effect of the programme from other factors, by setting control and treatment groups. Several experimental designs, such as one-group pre-test-post-test design, static group's comparison, pre-test, post-test control group design, Solomon four-group design, longitudinal study design, etc. are available in literature and can be used. However, in actual practice, extension programmes are seldom run in a way that allows an experimental design of evaluation. In Pilot Projects, it might be possible to use an experimental design of evaluation. By and large, a survey method is use. This method can be used for evaluating ongoing progress or as an ex-post facto evaluation of the programme after it has completed its tenure.

(vii) Collection and analysis of evaluation evidence

There are many methods for collecting information for evaluative purposes, such as the mail questionnaire, personal interview, distributed questionnaires, group interviews, case studies, systematic field observations, systematic study of secondary data etc. Selection of the right kind of data collection method will depend on the objectives of the evaluation, kind of information needed, time and resources available and the type of respondents from whom information is to be collected.

However, whatever the method used a specific questionnaire or interview schedule or data recording sheet must be developed with care. Once the data is collected, it must be tabulated, summarized and analyzed with adequate care. This step should not be rushed. To avoid delay, however, analysis may be done with the help of a computer.

(viii) Interpretation of the results in a proper way

It is a very crucial as evaluation results can be missed also. Once tentative generalizations are arrived at, it may be appropriate and they are informally discussed among the interpreters as well as with programme planning and implementation officials, so that the results of evaluation are put in a proper perspective.

The evaluation results must clearly state the achievements, failures and future adjustments needed. A written report of the evaluation findings should be prepared and made available to all concerned.

Important question related topic

- Evaluation word origin from Latin word *"Valere"* which means to be strong or valiant.
- Evaluation became an integral part of the extension education process.
- Describe the evaluation of extension programme.

■■■■

6

Extension Systems in India

6.1 EXTENSION EFFORTS IN PRE-INDEPENDENCE ERA

In the 19th Century, many efforts were held to end the difficulty of rural loaning system. The tradition of *'Takabi'* loan system' (advance of money given for cultivation, for digging wells, bringing improved seeds etc.) which was started from the time of Mughal emperors, continued during the British period and was further strengthened. But, above all it was too little to improve the conditions of rural people.

Similarly in the 19th Century some efforts were undertaken by establishing the rural banks by state governments, Justice Rana De and William Bad Khoran. The churchman Colson, who was associated with this planning, experienced that not only deposited capital is needed, but also the feeling of frugality and *"Self-help"* should be created in the minds of people. Unfortunately, this thinking could not be continued. In 1947 before achieving freedom in India, many programmes were started which are mainly as follows:

6.1.1 Sriniketan

Sriniketan is situated about 100 km. away from Calcutta (now Kolkata) in West-Bengal State. This area was socially, economically and politically backward. Shri Ravindra Nath Tagore began this project of village development in this area with the help of sociologist Shri L.M. Hurst. Shri Tagore thought that if some villages were developed, the other villages will get inspiration and the programme of village development will spread all over the country and thus the whole country would be developed. Sriniketan Village Welfare Institute was established in 1920 whose main objectives were the following:

1. To increase the knowledge of rural people.
2. To help the rural people in establishing cottage industry.
3. To inspire the people to follow new technology.
4. Development of dairy farming.
5. To create the feeling of co-operation.
6. To arrange the facility of health and education.
7. To create the feeling of rural leadership.

This Project, in the beginning, was started only in 8 villages, but after some time this project was extended to 15 villages. Shri Tagore was the landlord of all these villages. Being centralized over Shri Tagore, this project could not survive after the death of Shri Tagore.

6.1.2 Marthandam

This project of rural development was initiated in the village Marthandam near Trivandrum of Kerala State by Young Men Christian Association (YMCA) and Christian Church Association (CCA) under the direction of Dr. Spencer Hatch in 1928. This village was undeveloped economically and the economic condition of the native majority was poor. Here, people used to cultivate only paddy and coconut on some places. For exploiting this weakness, it was thought that some developmental work should be done, so that the Christian faith could spread. Consequently Dr. Hatch made agreement with Y.M.C.A. and the Church for his work and initiated this project in neighbouring village Marthandam.

The objectives of this project were also made on the basis as to how the public of this area could become Christian. The main objectives of this project were:

Spiritual Development – To develop the sentiments of people or to create the feeling of progress in their minds, because a work cannot be successful until the person is confident that the work will definitely be successful. Dr. Hatch had the confidence that if people are inspired for progress, then people can be motivated for these development works.

Mental Development – To just awaken the feeling is not enough? Mental development is also necessary, until the thinking, understanding and the capacity of taking decision is not developed, the work cannot be completed. Thus mental development of a man is also necessary.

Physical Development – The individual, who is not physically strong, cannot work well, whether he is developed mentally and spiritually. So, the physical development of a man is also necessary.

Social Development – If the person's social environment is adverse, he will be afraid to do any new work. Since all the activities of a person is determined by the society and so the development of their society is essential. Thus, the society in which a person lives, its development is also necessary.

Economic Development – Even if the man is spiritually, mentally, physically and socially developed, he will not work if he has no hope of economic advantage. So, the fifth objective of this project is to give economic advantage to the people. The main programme planning of this project were the following:

1. Before launching the programme, surveys are made to know the needs of that area and on the basis of the surveys the programmes were introduced according to the needs of the people.
2. The rural drama, rural exhibition, inter-rural competition, demonstration were also organised to affect the people and to attract the people.
3. Religious programmes were also organised for developing the religious feelings.
4. All-round development of rural life and each individual's progress were the subjects to importance in the progra-mmes, *i.e.*, Farming, rural industry, cooperation and to develop 'Panchayat' etc.
5. For bringing economic development among the rural people, many programmes as bee keeping, soap making and the educational programmes etc. were organised. This programme of Dr. Hatch remained for a long period because he had trained workers and the local people were inspired to work themselves. As such this project was very

successful in this area, but the whole programme was centred on a person and after the death of Dr. Hatch, there was a conflict on the question of leadership between YMCA and Church Association, and due to no economic advantage to the non-paid workers, this programme could not survive.

6. Baroda Village Reconstruction Project: Shri B.T. Krishnamachari in Baroda in the Gujarat State initiated this Project in 1932. This Project Continued for a long period.

Objectives

1. To improve the life style of rural people rapidly.
2. To spread education and industrialization.
3. To develop the necessary factors for the progress of agriculture.

Scope

1. This Project was started in the district of Navsari in the Gujarat State.
2. Many programmes such as gardening, poultry-farming, beekeeping. Spinning and weaving were organised.
3. Re-stabilization of 'Panchayats' and other programmes of village progress were organised.
4. The adult education had been extended.

Working System

1. By personal education and contact by the village guides.
2. To use the school teacher of village in the extension of programme.
3. To use the traditional means of extension.

6.1.3 Firka Development Scheme

The government of Madras (now Tamil Nadu) decided to make efforts for the development of villages at Firka level. The first programme began in 1946. Among Pre-Independence project, this was the biggest project.

Objectives

1. All-round development of rural people.
2. To develop the means of drinking water and sanitation.
3. To develop the committees of Panchayat and co-operatives.
4. To develop animal husbandry farming and facilitate irrigation.
5. To introduce khadi and cottage industry.

Scope

1. To select Firkas (villages) considering the possibilities for development in the production of handloom cloth and other cottage industries and alleviating backwardness.
2. This work was started from 34 to 84 Firkas till 1950.

6.1.4 Gurgaon Experiment

In Gurgaon district, this programme of village development was the first one to be run by the State it was started by the Mr. F.L. Brayne. In 1920, Mr. Brayne had been appointed on the post of Deputy Commissioner in Gurgaon district and he began this project of rural upliftment in his district, which became famous as "Gurgaon Project". The main objectives of this project were:

1. To increase crop production,
2. To control extra expenditure,
3. To improve the health,
4. To develop the feeling of women-education, and
5. Home development work.

Although this project got some success yet this scheme also could not survive more because this project was also based upon the sentiments of F.L. Brayne and when he was transferred, gradually this programme also stopped.

6.2 POST-INDEPENDENCE ERA

The urgent need for stepping up food production was realized even in the pre-Independence era and a Grow more food Campaign was started. Under the campaign, targets for increased agricultural production were laid down for the first time on an all-India basis. But the campaign failed to achieve its targets. Soon after Independence the Central Government re-defined the objectives of the Grow more food Campaign as the attainment of self-sufficiency in food grains by 1952, and simultaneously increased the targets of production of other crops to meet the short fall as a result of the partition of the country. At the same time, arrangements were made for integration and co-ordination of the entire campaign for increasing agricultural production. Some state governments associated the public with working of the campaign by setting up of non-official committees at the village, "*Taluka*", district and state levels. The plans were revised from time to time in order to make the campaign more effective. Accordingly, some initiatives in this direction were taken, which are given as under:

6.2.1 Etawah Pilot Project

This is such an effective project that after witnessing its results, the way was cleared for initiating the Community Development Project. Therefore, this was called the Pilot Project. In 1948, after Independence, the Government of India prioritized on rural development and how this work should be managed. For this, the guidance of an experienced person was needed. The Government of India urged the U.S.A. Government to send Mr. Albert Mayor to India because he had enough experience of rural development programme and was a Rural Sociologist. He also had the experience of Organising different programmes of village upliftment. Mayor, after arriving here, surveyed throughout India and on the basis of his experience, he concluded that it is a vast country enriched with different resources and seasons, therefore, such a field should be selected that can represent most parts of the country. So, Mr. Mayor made his headquarters at Maheva, District Etawah in U.P., selected 64 villages of neighbouring area and there, in 1948, initiated the work of development guided by the following objectives. Later this project guided '*Community Development Project*' in India; it was called *"Etawah Pilot Project"* because literally it is the pilot who shows the path.

Objectives

1. Mental development of the villagers.
2. To awaken the desire of rural people and to make them laborious.
3. The development of agriculture and animal husbandry.
4. The development of Panchayat.
5. To develop the feeling of self-confidence, co-operation and mass participation.
6. To seek the possibility of transferring this project elsewhere in the country.

Programme

1. To find out the desired needs of the people.
2. To demonstrate and evaluate the programmes.
3. To appoint the village level workers (VLW) for working as a link between the people and the government.
4. In agricultural economics, to develop the cottage industry as the co-operative business for development of the agriculture.
5. To open the stocks of seeds, fertilizers and the instruments of agriculture.
6. To introduce the people to the international world through news bulletins and newspapers.
7. To start the health programme for the rural people.
8. New schools of primary education and adult education for the extension of education were opened.
9. To solve the agricultural problems for the development of agriculture, *i.e.*, improvement of sterile land, to prevent the soil erosion, reconstruction of tube-wells, the arrangements of demonstrations for teaching them about improved agricultural methods, the construction of roads, the distribution of seeds and fertilizers and development of

horticulture etc. programmes of agricultural development were started.

Conclusion

From this Pilot Project, the Government of India achieved encouraging results. Inspired by the success of this project, some projects were introduced in Gorakhpur, Faizabad, etc. Government of India later on introduced a new programme of *"Community Development"* in 1952.

6.2.2 Nilokheri Experiment

In 1948, S.K. Dey prepared this project for the purpose of providing residence for 700 immigrants from Pakistan. He began this project using 100 acre of swampy land spreading in the midst of Karnal and Kurukshetra. The name of this project was *"Majdoor Manzil"*. The director of this project was Shri S.K. Dey. He went on to become the Union Minister of Community Development in 1965.

Objectives

1. Self-dependence in all the fields of life.
2. To arrange for professional training and provide occupation for the people on the basis of their experience.
3. To eliminate the middle-man system.

Scope

1. The work was done in 100 villages near Nilokheri.
2. To establish the village level workers, social development officer and block development officer.
3. The development and popularization of co-operative Institutions.
4. The facility for entertainment.
5. The effort to establish a socialist society.

Important points of the topic

- *"Takabi"* loan – Advance of money given to cultivation for digging wells, brining improved seeds.

- Sriniketan located – Away 100 kms from Calcutta in West Bengal
- Sriniketan area was backwards – Socially, Economically, Politically
- Project was started by – Ravindra Nath Tagore in 1920
- Project was started in – Eight villages after some time extended in 15 villages
- Sriniketan project based on concept – Brati Balika
- Marthandam project was started in – Trivandrum, Kerala
- Marthandam started with the help of – YMCA and CCA
- Marthandam project based on concept – Cooperation
- Pioneer person of Marthandam project – Spencer Hatch in 1928
- Main objectives of Marthandam – Spiritual, Mental, Physical, Economic and Social Development
- Village reconstruction started by – B.T. Krishnamachari
- Firka development schemes was started by – Madars (Tamil Nadu) Government
- Firka Development Schemes was started in – 1948
- Firka means – Village
- Work was started from – 34 to 84 Firka till 1950
- Gurgaon experiment was started by and in – F.L. Brayne, 1920
- Gurgaon experiment project based on – Village guide
- Brayne was – Deputy Commissioner of Gurgaon district
- Etawah project started in – 1948 by Albert Mayor
- Mayor made his headquarter at – Maheva village Etawah
- First time selected – 64 villages
- Pilot called because – Literally it is the pilot who shows the path
- Nilokheri started by and in – S.K. Dey, 1948

- Purpose of Nilokheri – Providing residence of 700 immigrants from Pakistan
- Area use – 100 kms of swampy land spreading in midst of Karnal and Kurukshetra
- Nilokheri also known as – Majdoor Manzil
- Explain any two pre and post independence programme

■■■■

7

Agriculture Development Programmes

7.1. INTENSIVE AGRICULTURE DEVELOPMENT PROGRAMME (IADP)

The major outcome of the above thinking was the formulation of a strategy of intensive approach to agricultural production, specially the food grains. A new programme named as IADP was formulated which was launched gradually from 1960. The third five year plan (1961–1966) incorporated this programme into the planned development process.

This programme was popularly known as a ***"Package programme"***. This name was given because of the collective and simultaneous application of all practices of improved seeds, irrigation, fertilizer, plant protection, implements, credit, etc.

This programme was started in July 1960 in seven selected districts in various states. They were (i) West Godavari in Andhra Pradesh, (ii) Shahabad in Bihar, (iii) Tanjore in Tamil Nadu, (iv) Raipur in Madhya Pradesh, (v) Ludhiana in Punjab; (vi) Pali in Rajasthan; and (vii) Aligarh in Uttar Pradesh.

The Criteria for Selection of Districts

1. Districts with high potentially for increasing the yield in short time, *i.e.,* with assured irrigation facilities and minimum natural hazards.
2. Districts with well-developed Panchayats and Cooperatives IADP is popularly known as **Package Programme**

because of the collective and simultaneous application of all improved packages of practices namely improved seed, irrigation, fertilizers, plant protection, implements, storage facilities, marketing facilities, credit etc.

Objectives

i. To achieve rapid increase in the level of agriculture production through a concentration of financial, technical, extension and administrative resources;

ii. To achieve a self-generating breakthrough in productivity and to raise the production potential by stimulating the human and physical process of change; and

iii. To demonstrate the most effective ways of increasing production and thus, to provide lessons for extending such intensified agricultural production programmes to other areas.

Weaknesses

- Educational approach to reach the cultivators was lacking.
- Poor trainings to staff.
- Staff was not clear about the methods to reach the cultivators.
- Posting of staff was not adequate and timely.
- Workshop, seed testing and soil testing laboratories were not functioning to the required level and transport and land development programmes were not progressing satisfactorily.

Achievements

In spite of these limitations IADP created a new dynamism in the farming community, greatly enhanced the use of technical inputs and also contributed to the growth of marketing and storage facilities as well as supporting services. Other achievements are:

1. Technical assistance was given to farmers in preparing production plans.

2. The cultivators were provided simultaneously with all supplies and services at right time and in adequate quantities through cooperatives
3. Credit was given to all who had their production plans and participated in the programme
4. Marketing and storage facilities were developed within bullock cart distance
5. Covered all the important cash crops grown in the districts all though emphasis was laid on the increase of food grain crops

7.2 INTENSIVE AGRICULTURE AREA PROGRAMME (IAAP)

Even with the short coming of IADP, the spectacular results in improved crop yield obtained in IADP, prompted the Government to think of extending the benefits of improved technology in agriculture in large areas over the country at less cost and with reduced staff strength. This resulted in launching of the IAAP in 1964. This was similar to but a less intensive extension programme in comparison to that of the IADP. 114 districts were selected in the year 1964 and later extended to 150 districts.

Achievements

- Achieved increased production by exploiting the land resources.
- Package approach were covered in 1410 communities blocks spread over 114 districts in India. In Karnataka 57 C.D. Blocks spread over 14 districts were covered.
- Increased production by 20–25 per cent of the cultivated area was achieved.
- Effective coordination between officials and nonofficial was achieved.
- Multiplication of improved seeds and its distribution to all cultivated areas was possible

Difference between IAAP and IADP

IADP	IAAP
The main objective of the IADP was to implement the improved farm practices in the selected potential areas to increase the production	The objective of the IAAP was to extend the concept of IADP to other potential areas to cover at least 20–25% of the cultivated area in the country.
This programme was recommended by a team of Ford Foundation experts	This programme was recommended by the third Five Year Plan mid? term appraisal committee.
The main criteria for selection of the districts was based on areas having assured irrigation facilities minimum natural hazards and well developed co-operatives	The main criteria for selection of districts was based on areas having predominant crops and well developed infrastructure facilities
Twenty eight districts including Mandya and Raichur were covered under this programme	One hundred and fifty districts including fourteen districts of Karnataka were covered under these programmes.
Number of personnel working in each community block were VLW's - 10 AEO - 1 SM's - 2 (SM's were responsible for the whole district)	Number of personnel working in each community block were VLW's - 10 AEO - 1 SM's - 2 (similar to IADP)

7.3 HIGH YIELDING VARIETIES PROGRAMME (HYVP)

HYVP is launched in 1966, which helped the country in attaining self-sufficiency in food. The technological development did not remain confined to the introduction of high yielding crop varieties alone. These were combined with the application of high analysis and balanced fertilizer, irrigation, plant protection, improved implements etc, which made a *'Green revolution'* possible in the country. The pervasive influence of high yielding technology spread to other areas of farm production such as animal production, fishery, sericulture, social forestry etc.

Punjab, Haryana and Western parts of Uttar Pradesh were initially selected for the phased launching of this strategy. The cultivation of HYV since 1966–67 had resulted in a substantial increase in food grains production. Wheat production was doubled. Rice production also had a substantial increase, though not as much as in the case of wheat.

The target of coverage of 2.5 crore hectares of area under HYVs of cereals and millets under fourth five year plan was exceeded. The coverage was more than four crore hectares.

7.4. KRISHI VIGYAN KENDRA (KVK)

The KVKs have witnessed several changes in their functions over the years. Accordingly their functional definition also has radically got refined so as to meet the new challenges in agriculture.

The most recent definition of a KVK is as follows:

"KVKs are grass root level organisations meant for application of technology through assessment, refinement and demonstration of proven technologies under different 'micro farming' situations in a district" (Das, 2007).

Role of KVKs in the Context of Agricultural Extension in India

Agricultural extension in India is largely deployed by government, implemented mainly through government institutions and to some extent through non-government agencies. Krishi Vigyan Kendras (KVKs) or Farm Science Centres as institutes of inducing behavioural change, are being managed by both government and non-government organisations. Literally, Krishi Vigyan Kendras have to serve as repository of scientific knowledge that is useful to the entire district, which is its jurisdiction. In India, agricultural extension and extension education are interchangeably used with the same connotation as used in American tradition, meaning "Extending Information" as a means of educating people to solve their problems. As a result, agricultural extension in India is more of "Informative Extension" than "Emancipatory Extension".

Recent-691, Dr. A.K. Singh, Deputy Director General (Agricultural Extension)

The first KVK was established in 1974 at Pondicherry under Tamil Nadu Agricultural University. The Krishi Vigyan Kendra is designed to impart need-based and skill-oriented vocational training to the practicing farmers, in-service field level extension workers, and to those who wish to go in for self employment.

The first KVK was established in 1974 at Pondicherry under Tamil Nadu Agricultural University. The priority for establishing KVKs is given to hilly areas, drought prone areas, forest areas, coastal areas, flood prone areas, and areas dominated with tribal farmers, weaker sections, small farmers and landless labourers. The objective is to gradually cover the entire country with one KVK in each district, priority being given to the backward areas.

The basic concepts of a KVK are

1. The centre will impart learning through work-experience and, hence, will be concerned with technical literacy, the acquisition of which does not necessarily require as a precondition the ability to read and write.
2. The centre will impart training only to those extension agents who are already employed or to practicing farmers and fishermen. In other words, these centres will cater to the needs of those who are already employed, or those who wish to be self-employed.
3. There will be no uniform syllabus for a KVK. The syllabus and programme of each centre will be tailored according to the felt needs, natural resources and the potentials for agricultural growth in that particular area.

The three fundamental principles of KVK are

- Agricultural production as the prime goal
- Work-experience as the main method of imparting training and priority to weaker sections of the society.

The main idea is to influence the productivity to achieve social justice for the neediest and deserving weaker sections of the society like the tribal farmers, small and marginal farmers, agricultural labourers, drought and flood affected farmers, and so on.

Need-based training courses are designed for different types of clientele. Courses are based on the information received through family and village survey. No certificate or diploma is awarded irrespective of the duration of the courses. After the training, follow-up extension programmes are organised for converting the acquired skills of the trainees into practice. While designing the courses, the concepts of farming system are taken into account to make the enterprises commercially viable.

Mandate

The mandate of a KVK is unique for it and is determined on the basis of the most important needs of the clientele, their resources and constraints, and nature of the ecosystem. The success of a KVK is judged by the extent to which it fulfils obligations specified in the mandate.

1. On-farm testing on farmers fields of proven technologies in agriculture and allied fields.
2. Organising Vocational Trainings in agriculture and allied areas
3. Conducting frontline demonstrations on major cereal, oilseeds, pulses and other important crops
4. Organising in-service training programmes to field/local extension functionaries in emerging advances in agriculture and allied areas.

The KVKs are fully funded by the Indian Council of Agricultural Research (ICAR). Initially, one KVK for each district was thought of and now two KVKs are also established in certain districts being the larger ones. Though KVKS are sponsored by ICAR they are working under different administrative controls *viz.*, SAUs, NGOs and ICAR.

7.5. INSTITUTION VILLAGE LINKAGE PROGRAMME (IVLP-195-96)

IVLP is an innovative program developed by the ICAR to help scientists to have direct interaction with the farming community so that appropriate technologies are developed for farmers. Here research, extension and farmers establish firm links by carrying together the assessment and refinement functions in the technology development and dissemination process. This helps the research system to generate a cafeteria of technologies, which are more productive in small production system, more profitable in commercial production system and gender sensitive for removal of drudgery of farmwomen.

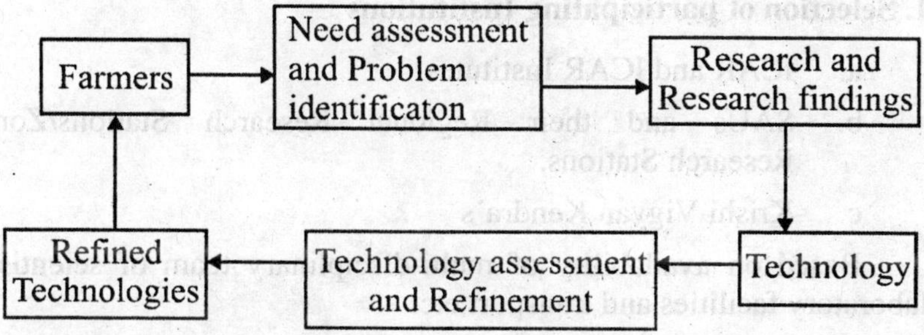

Research and Extension as an Integral Part of Technology Development

Objectives of IVLP as per ICAR guidelines are as follows:

1. To introduce technological intervention with emphasis on stability and sustainability along with productivity of small production systems.
2. To introduce and integrate the appropriate technologies to sustain technological interventions and their integration to maintain productivity and profitability taking environmental issues into consideration in a comparatively well defined production system.
3. To introduce and integrate the appropriate technologies to increase the agricultural productivity with marketable surplus in commercial on and off-farm production systems.

4. To facilitate adoption of appropriate post-harvest technologies for conservation and on farm value addition to agricultural products, by-products and waste for greater economic dividend and national priorities.
5. To facilitate adoption of appropriate technologies for removal of drudgery, increase efficiency and higher income of farm women.
6. To monitor socio-economic impact of the technological / technology modules based on environmental at meso and mega levels.

Methodology of implementation of IVLP as per ICAR guidelines are given below

1. Selection of participating Institutions

a. ICAR and ICAR Institutes
b. SAUs and their Regional Research Stations/Zonal Research Stations.
c. Krishi Vigyan Kendra's

Based on availability of multi-disciplinary team of scientists, laboratory facilities and transport etc.

2. Selection of village

One village or a cluster of villages to cover about 1000 farm families.

a. The selected village should not be far away from the research station
b. Should have asses through road
c. Should be a relatively poorly developed in agriculture

3. Agro-Eco-System analysis

Using Participatory Rural Appraisal Methods to gather information about

- The resource availability with the farmer's present production practices
- The extent of impact of the previous programs in the area

4. Constitution of multi-disciplinary team of scientists

Core team: Those scientists whose disciplines are essentially needed Optional team: The other disciplines depending upon the needs of the area.

5. Selection of team leader

6. Provide training to the multi-disciplinary team

7. Plan for Technology Assessment and Refinement

a. Intervention points based on problem – cause analysis

b. Basket approach considering the problems

c. Weightage to indigenous knowledge

8. Implementation of action plan

a. On-farm research

b. Demonstration

c. On-farm trials

9. Monitoring and Evaluation

a. Regular visit of team members

b. Technical staff posted for the village

c. IVLP card separately devised for the purpose

7.6. OPERATIONAL RESEARCH PROJECT (ORP)

The ORP was launched in 1975 by the ICAR. The basic aim of ORP was to demonstrate the technologies to the farmers and the extension workers on a watershed basis to evoke community action and participation

The main objectives of ORPs were

1. To test, adopt and demonstrate the new agricultural technologies in farmer's field in a cluster of three to four villages or in a watershed area.

2. To calculate profitability of the new technology meant for increasing production and economic returns substantially.
3. To identify socio-economic constraints affecting transfer of new technologies.
4. To assess the credit worthiness of the new agricultural practices.

The overall aim of the project was the socio-economic development of farmers with an integrated approach comprising of modern technologies of crops, horticulture, animal production, homestead, vocations and improving health hygiene and nutrition etc. by utilizing local available resources.

7.7 NATIONAL DEMONSTRATION (ND)

National Demonstration is a programme based on the concept of increasing the productivity per unit area and time by using proven agricultural technology. ICAR's National Demonstration programme on major food crops was launched in 1964.

The basic purpose of programme was to show the genetic production potentiality of new technology of major crops per unit of land and per unit of time and to encourage the farmers to adopt and popularize the technologies for accelerating

The main objective of this programmes are

1. To demonstrate convincingly to farmers, the production potentialities of a unit area of the land by using high yielding varieties of crops and adopting a multiple cropping programme with full package of practices such as balanced use of fertilizers and effective water management techniques.
2. To demonstrate use of implement for different operations and use of soil testing laboratories for use of balanced fertilizer doses.
3. To fully exploit these demonstrations for the purpose of training farmers in improved cultivation practices and to use them as recognized and effective audio visual aids for the flow of latest research technology and results to farmers.

4. To provide research workers a firsthand knowledge of the problems faced by farmers in growing high yielding varieties and to identify the constraints limiting the crop production.
5. To minimize the time lag between the research generated and its application in field. Production and improved cultivation practices.

The performance of high yielding varieties of cereals was very promising, but its full production potential was not demonstrated on a large scale to the farmers. Higher yields obtained on the research station did not impress the farmers. Who preferred to see the performance of these varieties on their own farms. Moreover, the scientists were looking for scientific feedback from field which would ultimately help them further refine the technology.

At this juncture in 1965–66 the ministry of Agriculture, Government of India initiated a nationwide programme in which demonstrations were connected on farmer's fields. This was the beginning of National Demonstration project (NDP). Demonstrations under this project were carried out mainly by the scientists of the SAU's and ICAR institutes in neighbouring villages. The scientists were required to demonstrate the potentiality of new seeds and package of practice on an area varying from 0.4 ha to 1.0 ha on farmers field. The single crop demonstration were carried out for crops like wheat, paddy, sorghum, pearl millet and maize.

Procedure to conduct national demonstration

- A demonstration has to be conducted on 0.4 ha land for full 1 year. Yield target of 9 tons of food grains for two crop per year per ha had been fixed. If achieved, the difference in the yield levels show the gap between how much is produced and how much can be produced, and how much can be produced by the farmers. A sum of Rs. 500 per year for three crops (Multiple demonstrations) was given for meeting the cost of critical inputs. For the first crop, the amount was Rs. 200 and for subsequent crops Rs. 150 each has been embarked.

- The NDP was implemented at district level by a team of 4 subject matter specialists. Who were entrusted with the responsibilities to conduct the demonstration on farmers field. They also organize field days at the time when important field operations were performed. Thus a large number of farmers got educated about new technology demonstrated on farmer's field.
- In all 25 demonstrations per season were laid on the farmers field in each district. Very intensive cropping system techniques of soil and water management including use of machinery and plant protection were demonstrated under National Demonstration Project (NDP). Thus these demonstrations served as pace settlers which were the first demonstrations and are to be emulated by local extension workers and farmers.

7.8 NATIONAL AGRICULTURAL TECHNOLOGY PROJECT (NATP- June, 30 1998)

The National Agricultural Technology Project (NATP) is a dynamic instrument of introducing major changes in the Agricultural Research and Extension systems of the country, besides developing their capabilities to meet future challenges. The project was initiated by Ministry of Agriculture, Govt. of India with the financial assistance of World Bank and would be implemented with the assistance of MANAGE in 28 districts covering 7 states, viz. Andhra Pradesh, Bihar, Jharkhand, Himachal Pradesh, Maharashtra, Orissa and Punjab over a period of 5 years (1998–2003). The World Bank assisted National Agricultural Technology Project (NATP) aims at improving research and extension services.

The Research component of NATP is being implemented by the Indian council of Agricultural Research (ICAR) and the Extension component by the Department of Agriculture and Co-operation. The different Project Implementing Agencies (PIAs) involved in implementation of the ITD component of NATP are:

i. Directorate of Extension (DOE)

ii. National Institute of Agricultural Extension Management (MANAGE)

iii. NATP cell at State Head Quarters

iv. State Agricultural Management and Extension Training Institute (SAMETI)

v. District Level Agricultural Technology Management Agencies (ATMAs)

The Extension component termed as "Innovations in Technology Dissemination"(ITD) envisages an integrated extension delivery at district level and is being pilot tested in seven participating states, *viz.,* Andhra Pradesh, Bihar, Jharkhand, Himachal Pradesh, Maharashtra, Orissa and Punjab.

Four districts in each of the seven participating states are identified for pilot testing as detailed below.

Objectives

The major objective of this component is

1. To accelerate the flow of technology form research, and extension to farmer.
2. Improve the dissemination of location specific and sustainability enhancing technologies.
3. Decentralize technical and decision making authority to the district level.
4. Create a more effective and financially sustainable public extension system.
5. Step up the privatization of certain technology transfer activities.

Salient Features

1. Pilot testing new institutional arrangements for technology dissemination at the district level and below through establishment of district Agricultural Technology Management Agency (ATMA).
2. Moving towards integrated extension delivery.
3. Bottom up planning procedures for setting the Research Extension agendas.
4. Addressing gender concern in agriculture and

5. Increasing use of information technology for effective dissemination.

7.9 NATIONAL AGRICULTURE INNOVATIVE PROGRAMME (NAIP)

The Government of India has launched the National Agricultural Innovation project with a credit support of the World Bank. The project will run up to June 2012. The ICAR is operating the Project. The overall objective of the project is to facilitate accelerated and sustainable transformation of Indian agriculture for rural poverty alleviation and income generation by the application of agricultural innovations through collaboration among public research organisations, farmers' groups, NGOs, the private sector and the civil societies and other stakeholders. The India National Agricultural Innovation Project contributes to the sustainable transformation of Indian agricultural sector to more of a market orientation to relieve poverty and improve income. The specific aim is to accelerate collaboration among public research organisations, farmers, the private sector and stakeholders in using agricultural innovations.

Agriculture is and will continue to be the main driver of country's economic growth with social justice. Our agriculture did extremely well and it was on the ascendancy till the mid nineties, but after that the growth slowed down. To assist the farmer in these changing contexts new strategies and innovative solutions are urgently required which in turn will require technological support. Hence the agricultural research system which generates technologies has to conduct the business of agricultural research in an innovative way. The World Bank aided National Agricultural Innovation Project (NAIP) has been conceived to pilot this innovation in conducting agricultural research.

The Basic Principles

1. To give the agricultural research and technology development system, an explicit development and business perspective through innovative models. In other words, the agricultural research system should be able to support agriculture as a business venture and also as a means of security of livelihood of the rural Indian while maintaining excellence in science.

2. To make the National Agricultural Research System a 'pluralistic' system where every Organisation having stake in agricultural research: public, private or civil society, has to play a role.
3. Working in well defined partnership groups with clear common goals and understanding on sharing responsibilities and benefits.
4. Funding through competition so that a wide choice of excellent innovative ideas come in from the stakeholders themselves.

Components of NAIP

The NAIP will function through four components:

1. The ICAR as the catalyzing agent for the management of change in the Indian NARS
2. Research on Production to Consumption Systems (PCS)
3. Research on Sustainable rural Livelihood security (SRLS)
4. Basic and Strategic Research in the frontier areas of agricultural sciences (BSR)

7.10 NATIONAL AGRICULTURE RESEARCH PROJECT (NARP)

The National Agricultural Research Project (NARP) was formulated by the Indian Council of Agricultural Research (ICAR) for strengthening the regional research capabilities of the state Agricultural Universities. Assistance for the project was provided by the International Development Association (IDA), an affiliate of the International Bank for Reconstruction and Development (IBRD). An agreement in this respect was signed in December 1978. The project was administered by the ICAR through a Project Funding Committee (PFC).

Objectives and approach of NARP

The main objective of NARP was to strengthen the regional research capabilities of the state Agricultural Universities as an important means of finding solutions to the location specific problems in different agro-climatic zones in their respective service areas.

For this purpose, intensification of research efforts was envisaged in respect of

1. Cereals, millets, pulses and oilseeds, particularly those that are grown under rainfed conditions,
2. Farming systems involving crop-livestock and crop-fish production systems
3. Agronomic practices
4. Soil and water conservation techniques potential

The objectives were sought to be achieved through

Rationalization of university research programmes and research organisations strengthening the infrastructural facilities of the state Agricultural Universities to undertake research on location specific problems

The main approaches for this comprise

a. Development/strengthening of at least one main station in each specific agro-climatic zone in the service area of the university, supported wherever necessary by sub-stations, by providing resource of staff, equipment and infrastructure needed to strengthen on-going research and to pursue new applied research.
b. Provision of resources, including infrastructure for verification of research results both at the main station and substations.
c. Provision of resources to initiate, strengthen and accelerate basic research on topics which are crucial for long term agricultural development of the state.

7.11 AGRICULTURAL TECHNOLOGY INFORMATION CENTRE

At present, there are 44 ATICs established in ICAR institutes and SAUs.

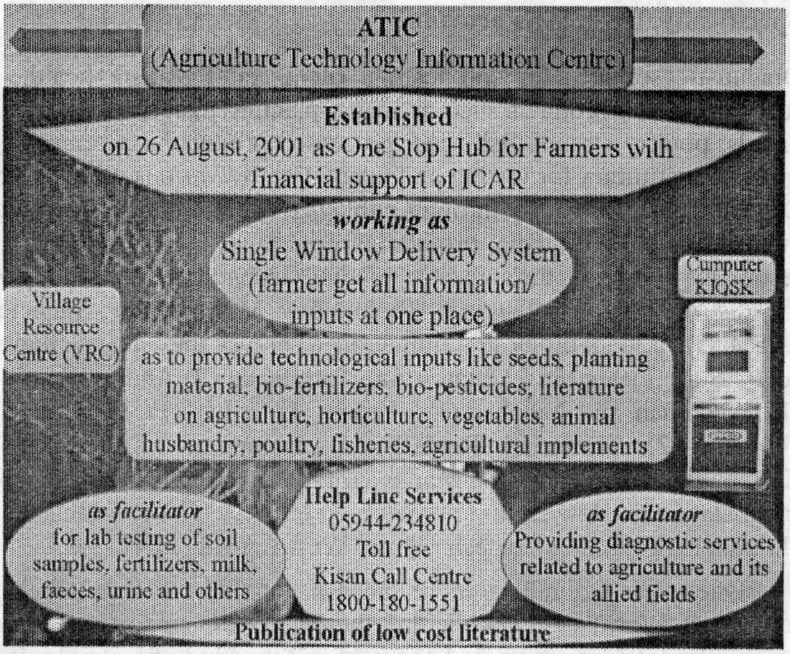

The Agricultural Technology Information Centres (ATICs) are established to provide greater coordination and intensive interaction between the researchers and technology users beyond individual units of research institutions in contributing towards the dissemination of information. Each ATIC will serve as a *"Single Window System"* with an objective to help Farmers and other stake holders such as Farmer-Entrepreneurs, Extension workers, Development agencies, Non-Government Agencies (NGOs) and private sector organisations to provide solutions to their location – specific problems in agriculture and make available all the technological information along with technology inputs and products for testing and use by them. The ATICs need to be demand driven and well integrated with research and financially sustainable, lay greater emphasis on location-specific and system based sustainable technologies. ATIC; ANGRAU started functioning from 1–1–1999 at Rajendranagar. ATIC disseminates technologies through: Office calls, Telephone calls, Personal letters, Expert node, Diagnostic services, Supply and Sale of technological inputs

Specifically, the ATICs provide

Diagnostic services for soil testing, plant and livestock health, a supply research products such as seeds, planting materials, livestock breeds, poultry strains, fish seed, processed products etc., emerging from an institution for testing and adaptation by various clientele. b. Disseminate information through published literature, audio - visual aids and electronic media, provide an opportunity to institutions for resource generation through sale of their technologies and support the district level Agricultural Technology Management Agencies (ATMAs) in technology dissemination wherever they are available. ATIC'S are located in 21 States viz., Tamilnadu (2) Andhra Pradesh (1), Andaman and Nicobar (1), Assam (1), Bihar (1), Gujarat (1), Haryana (2), Himachal Pradesh (3), Jammu & Kashmir (1), Karnataka (3), Kerala (5), Madhya Pradesh (3), Maharashtra (5), Meghalaya (1), New Delhi (1), Orissa (1), Punjab (1), Rajasthan (2), Uttaranchal (1), Uttar Pradesh (3) and West Bengal (1).

One of the key indicators of success of ATIC will be number of farmers and farmer groups visiting the centre, the sale of materials, the number of farmers trained per year and the kind of technological feedback received, documented and passed on to the research-extension systems.

7.12 RASTRIYA KRISHI VIKAS YOJANA (RKVY)

Launched - August 2007, **Closed-**2012

Sponsored by - Central government, funding pattern - 100% funded by centre, Ministry/Department - agriculture

Beneficiaries - Individual, family, community, women

Benefit type - Material, Loan and Subsidy

Eligibility criteria - To all the farmers in the states.

How to avail - Agriculture department (Nodal department) or other allied sector (Fisheries department, Horticulture, Animal Husbandry)

Background of RKVY

The RKVY aims at achieving 4 per cent annual growth in the agriculture sector during the XI Plan period, by ensuring a holistic development of Agriculture and allied sectors (Fisheries Department, Horticulture, Animal Husbandry etc).

Rashtriya Krishi Vikas Yojana is a special Additional Central Assistance Scheme which was launched in August 2007 to orient agricultural development strategies, to reaffirm its commitment to achieve 4 per cent annual growth in the agricultural sector during the 11th plan. The scheme was launched to incentivize the states to provide additional resources in their State Plans over and above their baseline expenditure to bridge critical gaps.

Concerned by the slow growth in the Agriculture and allied sectors, the National Development Council (NDC), in its meeting held on 29th May, 2007 resolved that a special Additional Central Assistance Scheme (RKVY) be launched. The NDC resolved that agricultural development strategies must be reoriented to meet the needs of farmers and called upon the Central and State governments to evolve a strategy to rejuvenate agriculture. The NDC reaffirmed its commitment to achieve 4 per cent annual growth in the agricultural sector during the 11th plan.

The Department of Agriculture, in compliance of the above resolution and in consultation with the Planning Commission, has prepared the guidelines for the RKVY scheme, to be known as National Agriculture Development Programme

Objectives of the programme

- To incentivize the states that increase their investment in agriculture and allied sectors
- To provide flexibility and autonomy to the states in planning and executing programmes for agriculture
- To ensure the preparation of Agriculture Plans for the districts and states
- To achieve the goal of reducing the yield gaps in important crops
- To maximize returns to the farmers
- To address the agriculture and allied sectors in an integrated manner

Basics features of RKVY

- It is a State Plan scheme
- The eligibility of a state for the RKVY is contingent upon the state maintaining or increasing the State Plan expenditure for Agricultural and Allied sectors
- The base line expenditure is determined, which is based on the average expenditure incurred by the State Government during the three years prior to the previous year.
- The preparation of the district and State Agriculture Plans is mandatory
- The scheme encourages convergence with other programmes such as NREGS.
- The pattern of funding is 100 per cent Central Government Grant.
- If the state lowers its investment in the subsequent years, and goes out of the RKVY basket, then the balance resources for completing the projects already commenced would have to be committed by the states.
- It is an incentive scheme, hence allocations are not automatic.
- It will integrate agriculture and allied sectors comprehensively.
- It will give high levels of flexibility to the states.
- Projects with definite time-lines are highly encouraged.

Areas of focus under the RKVY

- Integrated Development of Food crops, including coarse cereals, minor millets and pulses
- Agriculture Mechanization
- Soil Health and Productivity
- Development of Rainfed Farming Systems
- Integrated Pest Management
- Promoting extension services
- Horticulture

- Animal Husbandry, Dairying and Fisheries
- Sericulture
- Study tours of farmers
- Organic and Bio-fertilizers
- Innovative Schemes

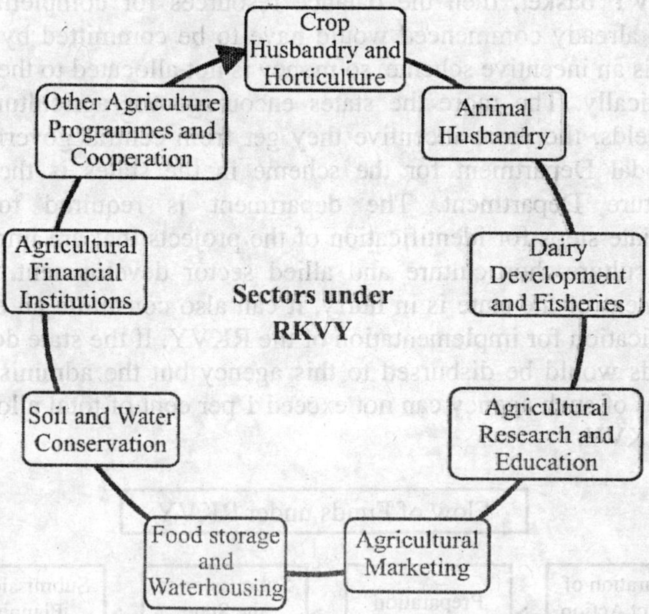

Incentivize the States RKVY

RKVY is a State Plan Scheme. How much assistance would be provided to a state from centre would depend upon the amount provided in State Plan Budgets for Agriculture and allied sectors, above a baseline expenditure on these sectors. This means that eligibility of a state for the RKVY is contingent upon the state maintaining or increasing the State Plan expenditure for Agriculture and Allied Sectors. The base line expenditure is determined, which is based on the average expenditure incurred by the State Government during the three years prior to the previous year. The RKVY funds are provided to the states as 100 per cent grant by the Central Government.

How states become eligible?

The states are mandatorily required to prepare the District and State Agriculture Plans that comprehensively cover resources and indicate definite action plans. States are encouraged to converge the scheme with other programmes such as NREGS, SGSY, BRGF, etc. If the state lowers its investment in the subsequent years, and goes out of the RKVY basket, then the balance resources for completing the projects already commenced would have to be committed by them. RKVY is an incentive scheme, so money is not allocated to the states automatically. The more the states encourage the agriculture and allied fields, the more incentive they get from central government. The Nodal Department for the scheme in the states is the State Agriculture Department. The department is required to take appropriate steps for identification of the projects that are important for agriculture, horticulture and allied sector development. If the government of the state is in hurry, it can also constitute an agency by notification for implementation of the RKVY. If the state does so, the funds would be disbursed to this agency but the administrative expenses of such agency can not exceed 1 per cent of total allocation under RKVY.

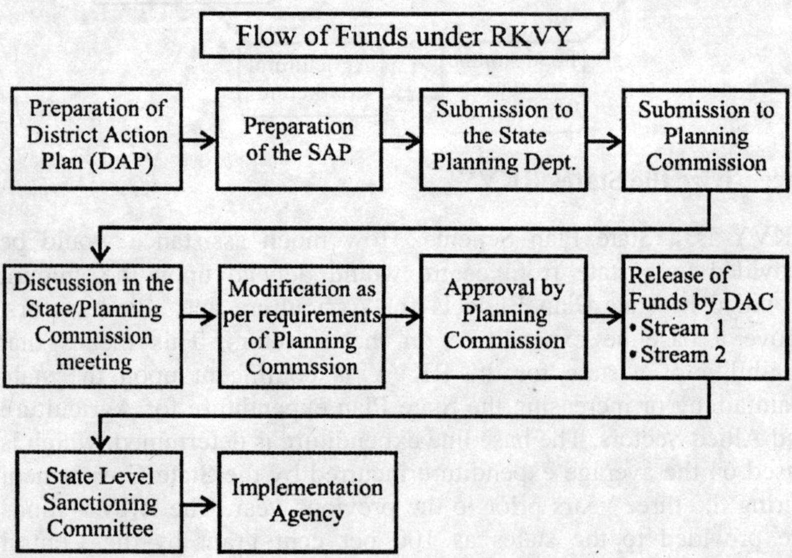

Composition of SLSC – Please note that the SLSC is chaired by the Chief Secretary of the state and its vice chairman is Principal secretary of agriculture in state departments. It has representations from the DAC, DAHD, and Planning Commission. The quorum for its meeting is incomplete without at least one GOI representative. It must meet in a quarter for at least once.

7.13 PRADHAN MANTRI FASAL BIMA YOJANA

Launched: 11 October, 2014

Main Objective: Provide insurance cover to Rabi and Kharif crops and financial support to farmers in case of damage of crops. In order to make crop insurance simpler and cheaper for the farmers, and to provide them with better insurance services, a Central Sector Scheme of Pradhan Mantri Fasal Bima Yojana (PMFSY) was launched by the Government of India replacing NAIS.

7.14 SOIL HEALTH CARD SCHEME

Launched: 17 February, 2015

Main Objective: To help farmers, to improve productivity from their farms by letting them know about nutrient/fertilizer requirements for their farms. The soil health card studies and reviews the health of soil or rather we can say a complete evaluation of the quality of soil right from its functional characteristics, to water and nutrients content and other biological properties. It will also contain corrective measures that a farmer should adopt to obtain a better yield.

7.15 NRLM

Aajeevika – National Rural Livelihoods Mission (NRLM) was launched by the Ministry of Rural Development (MORD), Government of India in June 2011. Aided in part through investment support by the World Bank, the Mission aims at creating efficient and effective institutional platforms of the rural poor, enabling them to increase household income through sustainable livelihood enhancements and improved access to financial services. NRLM set out with an agenda to cover 7 Crore rural poor households, across 600 districts, 6000 blocks, 2.5 lakh Gram Panchayats and 6 lakh villages in the country through self-managed Self Help Groups

(SHGs) and federated institutions and support them for livelihoods collectives in a period of 8–10 years.

In addition, the poor would be facilitated to achieve increased access to rights, entitlements and public services, diversified risk and better social indicators of empowerment. DAY-NRLM believes in harnessing the innate capabilities of the poor and complements them with capacities (information, knowledge, skills, tools, finance and collectivization) to participate in the growing economy of the country. In November 2015, the program was renamed Deendayal Antayodaya Yojana (DAY-NRLM)

NRLM Mission

To reduce poverty by enabling the poor households to access gainful self-employment and skilled wage employment opportunities, resulting in appreciable improvement in their livelihoods on a sustainable basis, through building strong grassroots institutions of the poor.

NRLM Guiding Principles

- Poor have a strong desire to come out of poverty, and they have innate capabilities
- Social mobilization and building strong institutions of the poor is critical for unleashing the innate capabilities of the poor.
- An external dedicated and sensitive support structure is required to induce the social mobilization, institution building and empowerment process.
- Facilitating knowledge dissemination, skill building, access to credit, access to marketing, and access to other livelihoods services underpins this upward mobility.

Implementation

NRLM is a highly process oriented programme and requires intensive application of resources, both financial and human, in order to mobilize the poor into functionally effective institutions, promote their financial inclusion and diversify and strengthen their livelihoods. It is, therefore, not feasible to roll out the programme in full scale across the country in one go, and therefore, it has been decided to phase the implementation of the programme over period of 10 years, blocks and districts in which all components of NRLM are implemented are treated as 'intensive' blocks and districts, whereas the remaining as 'non-intensive' blocks and districts.

Implementation at block level

NRLM intends to work in a block for a period of ten years till community federations take responsibility of implementation. A typical block having about 13,500 (90% of total poor) mobilizeable poor households spread over 100–120 villages is divided into 4 clusters of 30 villages each. In a typical intensive block, the first 3 years are spent in building the organisations of the poor by mobilizing them into SHGs, Federations at Village, Cluster level and Block level. Funds flow to the community institutions over the first 4–5 years. The middle (3–6) years, are invested in deepening the activities and addition of various layers, such as health, nutrition, interventions for persons with disability (PWD), etc. Last 4 years is essentially a maintenance and withdrawal phase where the community institutions graduate to self-reliance and self-sustainability.

Implementation in the blocks is being done in four ways –

a. **Resource Blocks** – With the support from National Resource Organisation (5–6% blocks in a state);
b. **Intensive Blocks** – Implemented with SRLM staff and internal community resource persons and the CRPs generated in resource blocks;
c. **Partnership Blocks** – With the support from local community federations and NGO partners;
d. **Non-intensive Blocks** – The remaining blocks in the state which are not taken up for implementation in the initial phase.

Support Structure

NRLM has set up dedicated sensitive support units at the National, State, District and Sub-District levels, to catalyze social mobilization, build institutions, capacities and skills, facilitate financial inclusion and access to financial services, support livelihoods and to promote convergence and partnerships with various programmes and stakeholders. These units would be staffed with professionally competent and dedicated human resources.

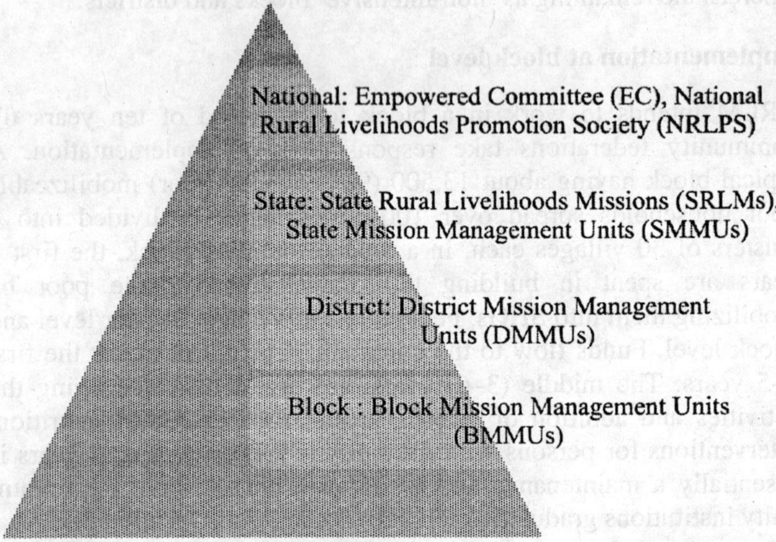

7.16 NATIONAL AGRICULTURAL HIGHER EDUCATION PROJECT (NAHEP)

The NAHEP is formulated by ICAR with a total cost of Rupees 1100 crores for four years starting from 2017–18. The project is proposed on a 50:50 cost sharing basis between the World Bank and the GOI, implemented at the Education Division, ICAR, New Delhi. NAHEP has been conceived to enable the agricultural education system catch up nationally and internationally with the peers, investing on infrastructure, competency and commitment of faculty, and attracting talented students to agriculture. It has been formulated with a focus to improve and sustain quality of higher agricultural education.

The project aims

To support infrastructure, faculty and student advancement, and providing means for better governance and management of agricultural universities, to raise the standard of current agricultural education system and provide more jobs and entrepreneurship orientation to agricultural education.

Project objectives

- Accentuate the relevance and quality of higher agricultural education in selected AUs
- Student and faculty development
- Improve learning outcomes, employability and entrepreneurship; and
- Enhancing institutional and system management effectiveness

Important points related to chapter

- Intensive Agriculture Development Programme was started in – **July, 1960**
- IADP was popularly known as – **'Package programme'**
- Package programme name was given because of – **Collective and simultaneous application of all practices**
- IADP was started in – **Seven districts various states**
- IADP first started in Rajasthan – **Pali district**
- Write short notes on IADP
- Intensive Agriculture Area Programme launched in – **1964**
- IADP was districts were selected in – **114 districts and later extended to 150 districts**
- HYVP was started in – **1966**
- Write short notes on HYVP
- KVK is – **Grass root level organisation**
- What is the KVK
- The first KVK was established in – **1974 at Pondicherry under TANU**
- KVK recommended by – **M.S. Mehta Committee**
- Write the mandates, objectives and principal of KVK
- Explain IVLP – **TAR with its methodology of implication**

- In IVLP – **One village cluster cover about 1000 farm families**
- Operational Research project was launched in and by – **1975 by the ICAR**
- ORP aim – **To demonstrate the technologies to the farmers and the extension worker**
- Write objective of the National Demonstration
- ND programme was started in and on – **1964 by ICAR on major crops**
- Describe NAIP and NATP
- What is the NAHE and NARP
- Explain ATIC with Structure
- RKVY was launched in – **August, 2007 by Ministry of Agriculture**
- Write objectives and features and sector under RKVY
- When was started Pradhsn Mantri Fasal Bima Yojana – **11 October, 2014**
- Soil Health Card Scheme started in – **17 February, 2015**
- National Rural Livelihood Mission in launched – **June, 2011, Ministry of Rural Development, GoI.**

■■■■

8

New Trends in Agriculture Extension

8.1 PRIVATIZATION EXTENSION

Process of funding and delivering the extension services by private individual or organisation is called private extension.

Savas (1987) privatization as: An act of reducing the role of government or increasing the role of private sector in an activity or in the ownership of assets.

Need of privatization

- Indian agriculture is self reliant in food production
- Low cost of technology

Concept of Privatization

Privatization of extension refers to services rendered in rural area and allied aspects of extension personnel working in private agencies or organisation for which farmers are expected to pay fee and it can be viewed as supplementary or alternatives to public extension services (Sarvanan and Shivalinge, 1980).

Privatization approaches

1. Share cropping system
2. Village extension contract system
3. Public extension through private delivery

4. Services for vouchers

Broad classification of Private Extension Services Providers

A. Private profit-oriented actors

- Consulting enterprise
- Processing/marketing enterprise
- Input supply enterprise
- Traders
- Private universities and research institutions

B. Private non-profit organisation

- NGOs
- Religious organisation

C. Producer Organisation

- Farmers association and federation
- Commodity grower and association
- Community and village organisations

Privatization models in different countries

New Zealand – Complete commercialization of public organisation

Netherland – Cost- recovery from farmers

China – Contracting of SMS by farmers groups

Ethiopia – Privatized service centres

Turkey – Cost-sharing of advisor

Kenya – Extension associated with contract out grower scheme

Private Extension Initiatives in India

Many of the private extension initiatives in India emerged without any active state support. Quite often they emerged in response to deficiency in public extension service provision.

The private extension service providers in case of agriculture extension include the following:

1. Input agencies (dealing with seeds, fertilizers, pesticides, equipments).
2. Large agri-business firms (involved in manufacture and sale of inputs and purchase of outputs).
3. Farmer organisations and producer co-operatives.
4. Non-governmental organisations (NGOs).
5. Media (print, radio and television) and web based knowledge providers.
6. Financial agencies involved in rural credit delivery.
7. Consultancy services.

The need for improved and expanded extension activities, together with a strengthening philosophical view of less government involvement in national economies, has led to a number of strategies for changing the way extension services are delivered.

Revitalization

The United States Cooperative Extension Service, when criticised for lack of relevance and vision (Dillman, 1986), regrouped and reviewed the criticisms. Its Extension Committee on Organisation and Policy (ECOP) organised a Futures Task Force to review issues and put forward recommendations with a view to revitalizing the system (ECOP, 1987), which has led to various alterations structurally and programmatically. Meanwhile, the advancement of electronic information systems is resulting in increased privatisation, with important implications for the future structure of U.S. Agriculture.

Commercialisation

New Zealand's Ministry of Agriculture and Fisheries' (MAF) agricultural advisory service now operates under user-pay, commercial criteria (Hercus, 1991). The MAF advisory service, renamed MAF Consulting and, subsequently. Agriculture New Zealand, has remained (temporarily) a public agency, although its employees have given up a number of public employment benefits and now receive commissions for consulting work undertaken. The agency depends for its annual budget on consulting fees received from farmers and contractual arrangements with government for the supply of policy information and rural intelligence to government.

Cost Recovery

Other public extension systems have moved toward cost-recovery approaches. Mexico has developed a fee-based system among large-scale farmers in the north-west region and plans the development of a similar arrangement among small-scale farmers in the south central region (Wilson, 1991). The Agricultural Development and Advisory Service (ADAS) in England and Wales, notionally "commercialized," operates on a partial cost-recovery basis. Clients of ADAS pay a fee for advice which formerly was free of charge. This process of cost recovery, introduced in 1987, was directed towards the agency receiving 50 per cent of its income from commercial fees by 1993–94 (Bunney and Bawcutt, 1991; Harter, 1992).

Voucher Systems

Some countries have replaced public extension delivery systems with vouchers, distributed by government services, for farmers to use in hiring private extension consultants (as in Chile). Coupons attached to agricultural bank loans, committing a certain percentage of the loan for extension services, have been used in Colombia.

Gradual "Privatisation"

In 1990 The Netherlands "Privatised" approximately one-half of its public extension service by transferring field extension personnel, with initial government financial support, to the farmer associations. The elements of the extension service responsible for linking research and the privatised extension services, policy preparation, implementation, and promotion and regulatory tasks remained under the aegis of the Ministry of Agriculture (Le Gouis, 1991). The "Privatised" extension service is governed by a board on which farmers' organisations and the government are equally represented (Proost and Röling, 1991).

Alternative funding and delivery

Diverse directions have been taken and multiple means of payment (public and private) have emerged as governments have opted for alternative financial and delivery arrangements to pay for and deliver public sector agricultural extension services. Extension provision is often multi-institutional and organised in ways that are not necessarily independent.

Where the public sector provides extension, the alternative funding arrangements include:

1. *General tax-based public funding* for agriculture, including funding of agricultural extension services, that is, the traditional public sector mode of funding extension
2. *Commodity tax-based public funding*, for example on an agricultural commodity such as coffee, as in El Salvador
3. *Fee-based public funding,* in which fees are charged, usually to large farmers for extension service, for instance in Mexico's grain rich northern region
4. *Contract-based commercialisation of public services,* whereby contract-based arrangements are made between farmer and public sector extension services, as in New Zealand

Private extension services are primarily of two types

- The first is the entirely private type which is directly involved in farming activities through consultants, agri-business, agricultural input industries etc.
- The second type consists of farmer's organisations, NGO's etc., which remain largely dependent on government subsidies.

Problems in public extension system

- Basic thinking is supply driven rather than demand driven,
- Limitedly focused on farmers' aspirations,
- Lack of location specific extension services,

- Extension services provided are general in nature,
- High cost, low impact of extension programmes,
- Poor motivation of staff working at grass root level,
- Curative nature of extension services,
- Inadequate technical qualifications of VLW,
- Incomplete extension services,
- Inadequate internal organisation structure,
- Inefficiency of extension personnel,
- Inappropriateness or irrelevance of extension content,
- Dilution of impact.

The private extension system in India offers the following services for farmers – terms of sharing, augmenting and supplementing the public extension efforts besides offering unique and innovative initiatives, which the public extension service can also emulate. Some of the Services are:

1. Cost sharing by farmers' groups
2. Cost recovery on selected services offered to farmers
3. Contracting services to small groups
4. Paid extension services for affordable farmers
5. Value addition by agro-processing firms
6. Consultancy services (both technical and managerial)
7. Privatised service centres for farmers
8. Self Help Groups of farmers
9. Information support through media organisations

Lesson from private extension

Based on a review of private extension initiative in India (Sulaiman, 2003) has identified the following lesson on private extension:

- The private extension offers much scope for supplementing and complementing public sector extension.

- Crop/commodity focused extension of private sector though very useful, is narrow in one sense as they do not engage with other related issues such as farmers organisation development or those issues related sustainability as well as public sector extension.
- While public funding would remain important, the delivery of all kinds of services need not necessarily be through the public sector. Several of these services could be either contracted by the public sector to other private extension providers such as NGOs and private consultancy firms for delivering specific services in select regions and client groups or jointly funded and implemented by public and private agencies.
- Private organisation is not a substitute for public extension and there is a need for significant public funding for extension in the years to come.

8.2 CYBER EXTENSION/E-EXTENSION

Cyber extension is extension of agricultural development with the help of Information and Communication Technologies over cyber space. Cyber space is an imaginary space behind networked computers through telecom means. This kind of a strong information sharing network is made possible through power of networks, computer communications and interactive multimedia.

Tools of Cyber Extension

As Cyber Extension means "Extension over cyber space", all the internet tools for developing and accessing Agricultural Information constitute the tools of Cyber Extension – 1. E-mail; 2. Expert systems providing information on pests and diseases; 3. Internet browsing for extension information; 4. Video conferencing; 5. Call centers and Satellite communication networks; 6. News and Discussion groups.

There are ICT initiatives for agricultural development in India many ICT projects supporting agricultural extension in India. A partial list of some famous ones is given in next page –

Project Name	Location
Gyandoot project	Madhya Pradesh
Warana Wired Village project	Maharashtra
iKisan	Andhra Pradesh
Automated Milk Collection Centres of Amul Dairy Cooperatives	Gujarat
Knowledge Network for Grass Root Innovations – Society for Research and Initiatives (SRISTI)	Gujarat
Land Record Computerization (Bhoomi)	Karnataka

8.3 MARKET-LED EXTENSION

Market led means identification of customers needs and wants before offering services

Objectives

After going through this unit you will be able to understand

- The need for Market-Led Extension
- The role of agricultural extension personal in market led extension
- The differentiate production led extension and market led extension

Challenges in agricultural marketing system

- Market size is large and continuously expanding, but marketing system not kept pace
- Private trade is 80 per cent marketed surplus
- Direct marketing "farmer – consumer" is negligible
- 85 per cent of the 27,294 rural periodic market, facilities for efficient trade is still almost absent
- 7161 market yards/sub yards is inadequate, ill equipped and mismanaged

- Due to lack of proper handling at farm gate lead to F&V (30%), grains (7%), spices (10%) loss before reaching market
- Rs 50,000 crores/year lost due to poor marketing chain
- Risk bearing: In both the production and marketing of produce, the possibility of incurring losses is always present. Market risks are those of adverse change in the value of the produce between the processes of production and consumption.
- Storage of farm produce: Whether storage takes place on the farm or in silos off the farm, increases in the value of products due to their time utility must be sufficient to compensate for costs at this stage, or else storage will not be profitable. These costs will include heating, lighting, chemical treatments, store management and labour, capital investment in storage and handling equipment, interest charges and opportunity costs relating to the capital tied up in stocks. Among the less tangible costs is the risks attached to storage. These include shrinkage due to pests, fungal growths and loss of quality due to ageing. Another risk is that demand could fall with adverse effects on prices.

Enhanced roles of Agricultural Extension personnel in light of Market-led Extension

SWOT analysis of the market: Strengths (demand, high marketability, good price etc.), Weaknesses (the reverse of the above), Opportunities (export to other places, appropriate time of selling etc.) and Threats (imports and perish ability of the products etc.) need to be analyzed about the markets. Accordingly, the farmers need to be made aware of this analysis for planning production and marketing.

- Organisation of Farmers' Interest Groups (FIGs) on commodity basis and building their capabilities with regard to management of their farm enterprise.
- Supporting and enhancing the capacities of locally established groups under various schemes/programmers like watershed committees, users groups, SHGs, water users' associations, thrift and credit groups. These groups need to

be educated on the importance, utility and benefit of self-help action.

- Enhancing the interactive and communication skills of the farmers to exchange their views with customers and other market forces (middlemen) for getting feedback and gain the bargaining during direct marketing ex. Rythu Bazars, Agri-mandi and Uzavar Santhaigal etc Establishing marketing and agro-processing linkages between farmers' groups, markets and private processors
- Advice on product planning: selection of crops to be grown and varieties suiting the land holding and marketability of produce will be the starting point of agri-enterprise. Extension system plays an important role in providing information in this regard
- Educating the farming community: to treat agriculture as an entrepreneurial activity and accordingly plan various phases of crop production and marketing
- Direct marketing: farmers need to be informed about the benefits of direct marketing. In some of the states, Rytu Bazars in AP, Apni Mandis in Punjab and Haryana and Uzavar Santhaigal in Tamilnadu have shown success in capacity building of FIGs in terms of improved production, post harvest operations, storage and transport and marketing
- Acquiring complete market intelligence regularly on various aspects of markets
- Regular usage of internet facility through computers to get updated on market intelligence
- Publication of agricultural market information in news papers, radio and television besides internet
- Organisation of study tours of FIGS: to the successful farmers/FIGs for various operations with similar socio-economic and farming systems as the farmers learn more from each other
- Production of video films of success stories of commodity specific farmers

Required information to extension system and farmers

- Present agricultural scenario and land use pattern
- Suitability of land holding to various crops/enterprises
- Crops in demand in near future

Market prices of crops

- Availability of inputs
- Usage of inputs
- Credit facilities
- Desired qualities of the products by consumers
- Market network of the local area and the price differences in various markets
- Network of storage and warehouse facilities available
- Transport facilities
- Regular updating of market intelligence

Production technologies

Like improved varieties, organic farming, usage of bio-fertilizers and bio-pesticides, IPM, INM, and right methods of harvesting etc.

- Post-harvest management like processing, grading, standardization of produce, value addition, packaging, storage, certification, etc. with reference to food grains, fruits and vegetables, eggs, poultry, fish, etc.
- Contract farming
- Private modern terminal markets
- Food retail chains
- Food safety and quality standard

8.4 FARMER-LED EXTENSION

Paradigm shift from Production-led Extension to Market-led Extension

Aspects	Production-led extension	Market-led extension
Purpose/ objective	Transfer of production technologies	Enabling farmers to get optimum returns out of the enterprise
Expected end results	Delivery of messages Adoption of package of practices by most of the farmers	High returns
Farmers seen as	Progressive farmer High producer	Farmer as an entrepreneur "Agripreneur"
Focus	Production / yields "Seed to seed"	enterprise/High returns "Rupee to Rupee"
Technology	Fixed package recommended for an agro-climatic zone covering very huge area irrespective of different farming situations	Diverse baskets of package of practices suitable to local situations/farming systems
Extensionists' Interactions	Messages, Training, Motivating and Recommendations	Joint analysis of the issues Varied choices for adoption Consultancy
Linkages/ liaison	Research-Extension-Farmer	Research-Extension-Farmer extended by market linkages
Extensionists' role	Limited to delivery mode and feedback to research system	Enriched with market intelligence besides the TOT Function Establishment of marketing and agro-processing linkages between farmer groups, markets and processors
Contact with farmers	Individual	Farmers' Interest Groups Commodity Interest Groups/SHG's
Maintenance of Records	Not much importance as the focus was on production	Very important as agriculture viewed as an enterprise to understand the cost benefit ratio and the profits generated

Aspects	Production-led extension	Market-led extension
Information Technology Support	Emphasis on production technologies	Market intelligence includ-ing likely price trends, demand position, current prices, market practices, communi-cation net work, etc. besides production technologies

8.5 EXPERT SYSTEMS

Objectives

After going through this unit you will be able to

- Understand the concepts and importance of Expert Systems
- Understand the components and designing of computer-based Expert Systems
- Apply the Expert Systems in various fields of agricultural sector.

An ES is defined as *"a computer program designed to model the problem solving ability of a human expert"* (Durkin, 1994).

It is also defined as *"a system that uses human knowledge captured in a computer to solve problems that ordinarily require human expertise"*.

Components of Expert system

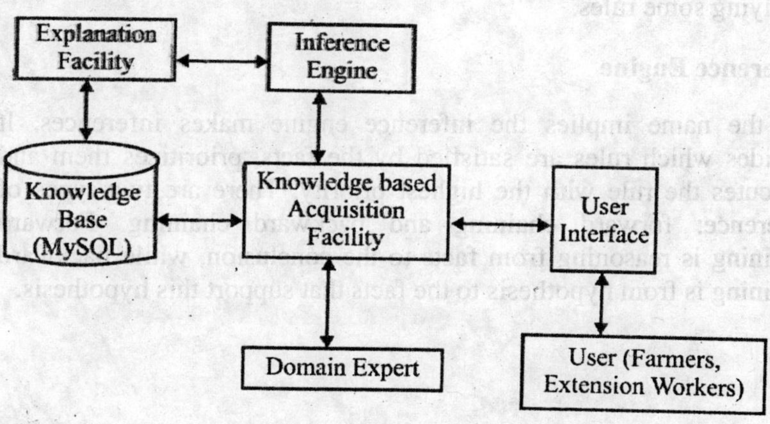

The ES consists of the following components:

User Interface

This is a mechanism to support communication between the user and system. The user interface may be a simple text-oriented display or a sophisticated, high resolution display. It is determined at the time of designing the system. Now-a-days graphical user interfaces are very common for their user-friendliness.

Explanation Facility

It explains the user about the reasoning process of the system. By keeping track of the rules that are fired, an explanation facility presents a chain of reasoning that led to a certain conclusion. So explanation facility is also called justifier. This feature makes a huge difference between ES and other conventional systems. Almost all the commercial ES shells do trace based explanation that is, explaining the inference on a specific input data set. Some systems explain the knowledge base itself and some explain the control strategy as well.

Working memory

This is a database used to store collection of facts which will be used later by the rules. More effort may go into the design and implementation of the user interface than in the ES knowledge base. Working memory is used by the inference engine to get facts and match them against the rules. The facts may be added to the working memory by applying some rules.

Inference Engine

As the name implies the inference engine makes inferences. It decides which rules are satisfied by the facts, prioritizes them and executes the rule with the highest priority. There are two types of inference: forward chaining and backward chaining. Forward chaining is reasoning from facts to the conclusion, while backward chaining is from hypothesis to the facts that support this hypothesis.

Knowledge Acquisition Facility

This allows the user to enter knowledge in the system thereby avoiding the need of knowledge engineer explicitly code the knowledge. It is an optional feature on many ES. Simple rules can be created using rule induction. In rule based ES, knowledge base is also called production memory as rules in the form of if-then are called productions. It is depicted in the picture as given below

Capabilities of Expert Systems

- Explore impact of strategic goals
- Impact of plans on resources
- Integrate general design principles and manufacturing limitations
- Provide advice on decisions
- Monitor quality and assist in finding solutions
- Look for causes and suggest solutions

Expert Systems Development Stages

The development stages are depicted diagrammatically as follows

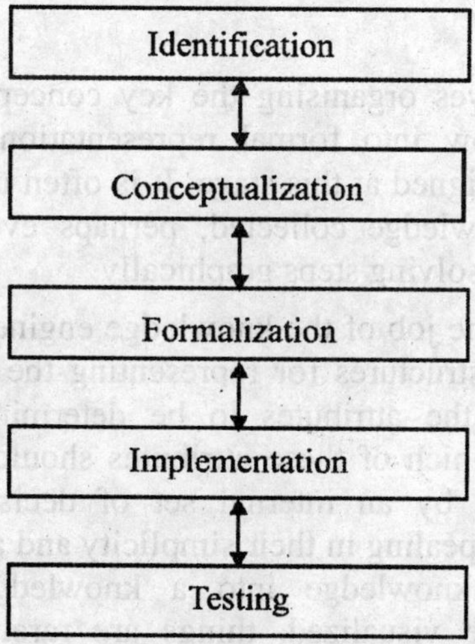

Identification

Identification is the requirements analysis step carried out in traditional software development. It involves a formal task analysis to determine the external requirements, form of the input and output, setting where the program will be used and determines the user. The participants, the problems, the objectives, the resources, the costs and the time frame need to be clearly identified at this stage. The participants are the group sponsoring the effort, the domain expert and the knowledge engineer. Choosing an appropriate domain expert is essential to the success of the project. The domain expert should be a legitimate authority in the subject matter area, as software must possess high quality knowledge, and this person must have time and interest to commit to the project.

Conceptualization

The second stage of ES development, conceptualization, involves designing the proposed program to ensure that specific interactions and relationships in the problem domain are understood and defined. The key concepts, relationships between objects and processes, and control mechanisms are determined. This is the initial stage of knowledge acquisition. It involves the specific characterization of the situation and determines the expertise needed for the solution of problem.

Formalization

Formalization involves organising the key concepts, sub-problems and information flow into formal representations. In effect, the program logic is designed at this stage. It is often useful to group or modularize the knowledge collected, perhaps even attempting to display the problem solving steps graphically.

In effect, it is the job of the knowledge engineers to build a set of interrelated tree structures for representing the knowledge base. They must decide the attributes to be determined to solve the problem and then which of these attributes should be asked of the user or represented by an internal set of decision trees. While decision trees are appealing in their simplicity and are a good way to begin formalizing knowledge into a knowledge representation scheme that can be visualized, things are rarely this simple in practice and rigid adherence to a tree structure is seldom satisfactory.

Implementation

During the next stage, implementation, the formalized knowledge is mapped or coded into the framework of the development tool to build a working prototype. The contents of knowledge structures, inference rules and control strategies established in the previous stages are organised into suitable format. Often, knowledge engineers will have been using the program development tool to build a working prototype to document and organise information collected during the formalization stage, so that implementation is completed at this point. If not, the notes from the earlier phases are coded at this time.

Consideration must be given to long-term maintenance. Modifications to the knowledge base over time must be anticipated. The knowledge base should be extensively documented as it is coded. The potential for later misunderstanding and confusion should be minimized wherever possible. Furthermore, extensive justifications and explanations should be included to assist the end-user in fully understanding questions posed to them by the program, so that the user can effectively use the program output, and to show the user, on demand, how the recommendation was logically derived.

Testing

This is the last stage which involves considerably more than finding and fixing syntax errors. It covers the verification of individual relationships, validation of program performance and evaluation of the utility of the software package. Testing guides reformulation of concepts, redesign of representations and other refinements. Verification and validation must occur during the entire development process. Verification proves that the models within the program are true relationships. It ensures that the knowledge is accurately mimicked by having the domain expert operate the program for all possible contingencies.

Advantages and Disadvantages of Expert System

Advantages of ES

The significant advantages in the above mentioned ES are given below

- The system can be used by extension personnel, researchers and farmers to identify crop diseases and enable to proceed their management.
- User can easily identify the disease on the basis of photographs of symptoms and text descriptions of disease.
- The user friendly software developed using window environment, thus provides enough facilities to identify the disease and to suggest the remedy conveniently.
- Provide consistent answers for repetitive decisions, processes and tasks.
- Hold and maintain significant levels of information.
- Reduce employee training costs.
- Centralize the decision making process.
- Create efficiencies and reduce the time needed to solve problems.
- Combine multiple human expert intelligences.
- Reduce the amount of human errors.
- Review transactions that human experts may overlook

Disadvantages of ES

- Lacks common sense needed in some decision making
- Cannot make creative responses as human expert would in unusual circumstances
- Domain experts not always able to explain their logic and reasoning
- Errors may occur in the knowledge base, and lead to wrong decisions
- Cannot adapt to changing environments, unless knowledge base is changed

Application of Expert system in the field of Agriculture

Mango ES

Prasad *et al.* (2006) formulated an ES *viz.,* AMRAPALIKA for diagnosing 14 different pests including eight diseases and six insects in Indian mango variety. The ES is developed for important diseases like Powdery mildew, Black spot, Anthracnose, Red rust, Die back, Bacterial spot, Sooty mould and Malformation, and insects like Shootborer, Red ants, White ants, Mealy bug, Mites and Fruit fly.

Rice ES

In India, Sharma *et al.* (2010) developed an ES in order to diagnose and manage the diseases occurring in rice crop. They have given an architectural framework of an ES in the area of agriculture and describe the design and development of the rule based ES, using the shell ESTA. The designed system is intended for the diagnosis of common diseases occurring in the rice plant. The rice ES is a computer program normally composed of a knowledge base, inference engine and user-interface.

Tomato Expert System

A web based tomato crop expert information system was developed by Babu *et al.* (2010) in India. The tomato crop expert advisory system is aimed at a collaborative venture with eminent agriculture scientists and experts in the area of tomato plantation with an excellent team of computer engineers, programmers and designers.

Rapeseed-Mustard Expert System

Vinod *et al.* (2008) developed an image based rapeseed-mustard disease expert system in India. The diagnosis and control measures of economically important diseases like alternaria blight, white rust and White rot, Downy mildew complex, Powdery mildew of rapeseed-mustard were effectively performed by using this expert system.

Wheat Pakistani

Khan *et al.* (2008) established a web-based ES for wheat crop in Pakistan. The rule-based ES covers two main classes of problems namely diseases and pests, normally encountered in wheat crop. This web-based ES shell allows JAVA interface to process its input and output sets.

Important questions

- Funding and delivering the extension services by private organisation-Private Extension
- Explain the classification of private extension services provider
- Name of Privatisation models in Turkey-Cost-Sharing of advisor
- Name of Privatisation models in Kenya-Extension associated with contract out grower scheme
- Name of Privatisation models in China-Contracting of SMS by farmers groups
- Name of Privatisation models in Ethiopia-Privatised service centres
- What are the problems in public extension system
- Extension of agricultural development with the help of information and communication technologies over cyber space-Cyber Extension
- Cyber extension means – 'Extension over cyber space'
- Gyandoot project initiated – Madhya Pradesh
- Warna Wired Village project – Maharashtra
- i-kisan – Andhra Pradesh
- Land Record Computerization (Bhoomi)–Karnataka
- What are the challenges in agricultural marketing system
- Differences between production–led extension and Market-led extension
- Write short notes on objectives of expert system
- Computer program designed to model the problem solving ability of a human expert – Expert System

- Components of Expert System – Five
- Write the capabilities of ES
- Explain the development stages of ES
- Write down the advantages and disadvantages of Expert system
- Describe the tomato and Rapeseed-Mustard Expert system

■■■■

9

Rural Development

9.1 CONCEPT, MEANING AND DEFINITION

Meaning and Definitions

The term rural development combines two words Rural and Development.

The term Rural and Development is used in different ways as given below.

- *As a Concept* – Development of Rural areas
- *As a phenomenon* – Interaction between institutional factors
- *As a Strategy* – Approach to bring positive change in rural life

Rural: According to International Encyclopedia of Social Sciences, there is a broad general consensus that the term rural refers emphatically to population living in the area of low density and to small settlements.

Development: "The term development, according to Lale (1975) is a process of improving standards of the masses of the low income population residing in rural areas making the process of rural development self-sustaining".

Definitions of Rural Development

According to Agarwal (1989), Rural Development is a strategy to improve the economic and social life of a specific group of people, the rural poor, including small and marginal farmers, tenants and landless.

According to World Bank rural development is a strategy designed to improve the economic and social life of a specific group of people by extending the benefits of development to the poorest among those who seek livelihood in the rural areas.

According to National Commission on Agriculture, rural development means development of an area and the people through optimum development and utilization of local resources by bringing about necessary institutions, structures and attitudinal changes, and by delivering package of services to improve all fields of the rural poor and rural weak.

The term rural development is a subset of the broader term 'Development'. However we define it, development is a universally cherished goal of individuals, families, communities and nations all over the world. Development is also natural in the sense that all forms of life on Planet Earth have an inherent urge to survive and develop. Given these two attributes, *i.e.*, its universal supremacy as a goal and its natural occurrence, development deserves a scientific study and analysis. Hence, it is not surprising that the subject of development has been studied by scholars of all faiths, ideologies and disciplines.

Concepts of Rural Development

According to World Bank (1975) – The rural development in general terms, is a strategy designed to improve economic and social life of people in a rural settlement and in particular, it focuses attention on the rural poor comprising the small and marginal farmers, tenants and landless labourers.

Rural development is the dynamic process of development of the rural people through various programmes and projects so that they can become self-reliant citizens of the country. The work is done by involving various agencies and organisations, and above all, the local people themselves.

It involves extending the benefits of development to the poorest among those who seek a livelihood in the rural areas. The group includes small scale farmers, tenants and the landless. As a phenomenon, rural development is the end result of interaction between various physical, technological, economic, socio-cultural and institutional factors. Motivate the people for adoption. As a strategy, it is designed to improve the economic and social wellbeing of a specific group of people – the rural poor. As a discipline, it is multidisciplinary in nature, representing an interaction of agricultural, social, behavioural, engineering and management sciences.

Robert Chambers (1983) defined that "Rural development is a strategy to enable a specific group of people, poor rural women and men, to gain for themselves and their children more of what they want and need. It involves helping the poorest among those who seek a livelihood in the rural areas to demand and control more of the benefits of rural development. The group includes small-scale farmers, tenants and landless."

Rural Development is a process of developing and utilizing natural and human resources, technologies, infrastructural facilities, institutions and organisations, government policies and programmes to encourage and speed up economic growth in rural areas, to create jobs and to improve the quality of rural life towards self-sustenance. Ultimate objective of rural development is improving the quality of life of rural poor and the rural weak.

Objectives of Rural Development

1. Providing goods and services in terms of social and economic infrastructure
2. Increasing the income of every rural family on a self-sustaining basis
3. Creation of additional employment opportunities in rural areas
4. It implies a broad based reorganisation and mobilization of the rural masses so as to enhance their capacity to cope effectively with the daily tasks of their lives and with changes consequent upon this
5. Improvement of services or rural masses in the process

6. Improvement of know-how, which is to be implemented to the rural People

According to Singh (1998), the main objectives of rural development in all societies, irrespective of their economic, political and socio-cultural systems are:

- To make available and improve the distribution of life-sustaining goods, such as food, clothes, shelter, health and security.
- To raise per capita purchasing power and improve its distribution by providing better education, productive and remunerative jobs and cultural amenities.
- To expand the range of economic and social choices to individuals by freeing them from servitude and dependence.

Problems in Rural Development

1. Most people are illiterate – for such people, extension teaching methods like Demonstrations, individual and group approaches, Training classes require large number of extension workers.
2. Inadequate communication channels especially Mass Media in rural areas
3. Limitation of funds and staff for training the farmers
4. As a traditional society with old ways and practices does not want to take risk unless they see the results
5. In an illiterate traditional society real leadership could not come forward
6. Communities and individuals differ in their needs as their circumstances change
7. Organisational constraints
8. Vaguely framed objectives of organisation

Policies and Strategies of Rural Development

The present strategy of rural development mainly focuses on poverty alleviation, better livelihood opportunities, provision of basic amenities and infrastructure facilities through innovative programmes of wage and self-employment. The above goals will be achieved by various programme support being implemented creating partnership with communities, non-governmental organisations, community based organisations, institutions and industrial establishments, while the Department of Rural Development will provide logistic support both on technical and administrative side for programme implementation. Other aspects that will ultimately lead to transformation of rural life are also being emphasized simultaneously.

The Government's policy and programmes have laid emphasis on poverty alleviation, generation of employment and income opportunities and provision of infrastructure and basic facilities to meet the needs of rural poor. For realizing these objectives, self-employment and wage employment programmes continued to pervade in one form or other.

As a measure to strengthen the grass root level democracy, the Government is constantly endeavouring to empower Panchayat Raj Institutions in terms of functions, powers and finance. Gram sabha, NGOs, Self-Help Groups and PRIs have been accorded adequate role to make participatory democracy meaningful and effective.

The prime goal of rural development is to improve the quality of life of the rural people by alleviating poverty through the instrument of self-employment and wage employment programmes, by providing community infrastructure facilities, such as drinking water, electricity, road connectivity, health facilities, rural housing and education and promoting Decentralisation of powers to strengthen the Panchayat Raj institutions.

The Department of Rural Development and Panchayat Raj is responsible for the implementation of various rural welfare schemes and also assists Panchayat Raj Institutions to discharge their duties and functions as effective Local Self Government entities.

9.2 VARIOUS RURAL DEVELOPMENT PROGRAMMES LAUNCHED BY GOVT. OF INDIA

9.2.1 Integrated Child Development Scheme (ICDS)

Launched on 2nd October 1975, today, ICDS Scheme represents one of the world's largest and most unique programmes for early childhood development. ICDS is the foremost symbol of India's commitment to her children – India's response to the challenge of providing pre-school education on one hand and breaking the vicious cycle of malnutrition, morbidity, reduced learning capacity and mortality, on the other.

1. Objectives: The Integrated Child Development Services (ICDS) Scheme was launched in 1975 with the following objectives:

i. To improve the nutritional and health status of children in the age-group 0–6 years;

ii. To lay the foundation for proper psychological, physical and social development of the child;

iii. To reduce the incidence of mortality, morbidity, malnutrition and school dropout;

iv. To achieve effective co-ordination of policy and implementation amongst the various departments to promote child development; and

v. To enhance the capability of the mother to look after the normal health and nutritional needs of the child through proper nutrition and health education.

2. Services: The above objectives are sought to be achieved through a package of services comprising:

i. Supplementary nutrition,

ii. Immunization,

iii. Health check-up,

iv. Referral services,

v. Pre-school non-formal education and

vi. Nutrition and health education.

2.1. Nutrition including Supplementary Nutrition: This includes supplementary feeding and growth monitoring; and prophylaxis against vitamin A deficiency and control of nutritional anaemia. All families in the community are surveyed, to identify children below the age of six and pregnant, and nursing mothers. They avail of supplementary feeding support for 300 days in a year. By providing supplementary feeding, the Anganwadi attempts to bridge the caloric gap between the national recommended and average intake of children, and women in low income and disadvantaged communities.

Growth Monitoring and nutrition surveillance are two important activities that are undertaken. Children below the age of three years of age are weighed once a month and children 3–6 years of age are weighed quarterly. Weight-for-age growth cards are maintained for all children below six years. This helps to detect growth faltering and helps in assessing nutritional status. Besides, severely malnourished children are given special supplementary feeding and referred to medical services.

2.2 Immunization: Immunization of pregnant women and infants protects children from six vaccine preventable diseases-poliomyelitis, diphtheria, pertussis, tetanus, tuberculosis and measles. These are major preventable causes of child mortality, disability, morbidity and related malnutrition. Immunization of pregnant women against tetanus also reduces maternal and neonatal mortality.

2.3 Health Check-ups: This includes health care of children less than six years of age, antenatal care of expectant mothers and postnatal care of nursing mothers. The various health services provided for children by Anganwadi workers and Primary Health Centre (PHC) staff, include regular health check-ups, recording of weight, immunization, management of malnutrition, treatment of diarrhoea, de-worming and distribution of simple medicines etc.

2.4 Referral Services: During health check-ups and growth monitoring, sick or malnourished children, in need of prompt medical attention, are referred to the Primary Health Centre or its sub-centre. The Anganwadi worker has also been oriented to detect disabilities in young children. She enlists all such cases in a special register and refers them to the medical officer of the Primary Health Centre/Sub-centre.

2.5 Non-formal Pre-School Education (PSE): The Non-formal Pre-school Education (PSE) component of the ICDS may well be considered the backbone of the ICDS programme, since all its services essentially converge at the Anganwadi – a village courtyard. Anganwadi Centre (AWC) – a village courtyard, is the main platform for delivering of these services. These AWCs have been set up in every village in the country. In pursuance of its commitment to the cause of India's Children, present government has decided to set up an AWC in every human habitation/settlement. As a result, total number of AWC would go up to almost 1.4 million. This is also the most joyful play-way daily activity, visibly sustained for three hours a day. It brings and keeps young children at the Anganwadi centre - an activity that motivates parents and communities. PSE, as envisaged in the ICDS, focuses on total development of the child, in the age up to six years, mainly from the underprivileged groups. Its programme for the three-to six years old children in the Anganwadi is directed towards providing and ensuring a natural, joyful and stimulating environment, with emphasis on necessary inputs for optimal growth and development. The early learning component of the ICDS is a significant input for providing a sound foundation for cumulative lifelong learning and development. It also contributes to the universalization of primary education, by providing to the child the necessary preparation for primary schooling and offering substitute care to younger siblings, thus freeing the older ones – especially girls to attend school.

2.6 Nutrition and Health Education: Nutrition, Health and Education (NHED) are a key element of the work of the Anganwadi worker. This forms part of BCC (Behaviour Change Communication) strategy. This has the long term goal of capacity-building of women, especially in the age group of 15–45 years, so that they can look after their own health, nutrition and development needs as well as that of their children and families.

9.2.2 Integrated Rural Development Programme (IRDP-1978)

Genesis

The CDP launched in 1952 could not bring expected change in improving the conditions of rural masses or there was not much achievement in the areas of agricultural development.

The apparent failure of the CDP was the main reason for the evolution of the IRDP. The main drawbacks of the CDP were

- Uneven distribution of the benefits of the programme.
- Absence of clearly defined priority in the programmes.
- The inability of the CDP to recognize and solve the inherent conflicts in the inter and intra target groups.
- Lack of mass participation.
- More bureaucratic.

Difference between CDP and IRDP

CDP	IRDP
Meant for total development of the Community *e.g.*, Road, drinking water	Meant for the individual development, who are below the poverty line?
Community approach	Family approach
Multi-dimensional and multi-sectorial programme	Integrated programme
The felt needs of the beneficiaries were not taken into account.	They are given due consideration
Financing is less intensive.	Financing is more intensive.

So, it was considered necessary to go in firstly for a programme directed mainly at agriculture and secondly to attempt the process of development in selected areas. Thus, in 1960 Intensive Agricultural District Programme (IADP) was formed and implemented. In this programme, efforts to boost agricultural production were concentrated in areas with better prospects of higher yields (package programme). The success achieved in increasing agricultural production under IADP, led the Government to extend the programme to other districts with slight modifications under the name of Intensive Agricultural area Programme (IADP) in 1964.

The food situation became alarming; particularly due to successive famines during 1064–67 and this happened despite intensive efforts in selected areas. In order to overcome this crisis, the GOI launched a new agricultural strategy known as High Yielding Varieties Programme (HYVP) in 1966. The achievement was the single goal of increasing agricultural production.

This unusual shift in the rural development strategy in favour of increasing agricultural production led to the accentuation of regional disparities on one hand and economic inequalities among different sections of the population on the other. Various reports revealed that the gains of intensive development efforts, including institutional credit flew more towards large and resourceful farmers to the neglect of small farmers, landless labourers, tenants and artisans.

On realization of the above, there was a shift in the rural development strategy. The main purpose of such a change in the approach was to make deliberate efforts to flow development benefits to the poorer sections and the backward areas. Accordingly, Small Farmers Development Agency (SFDA); Marginal Farmers and Agricultural Labourers (MFAL); Drought Prone Area Programme (DPAP); Hill Area Development Programme (HADP) etc., were implemented.

The beneficiary-oriented programmes as well as area development programmes, despite some successes could not make much dent into the problem of poverty and unemployment. Considering the magnitude and dimensions of rural poverty, the rural development strategy, therefore, took a major turn in its emphasis, content, coverage and methodology. This resulted in launching of the programme "Integrated Rural Development Programme (IRDP) in 1978–79, covering 2,300 development blocks and extended to all the 5011 blocks in 2nd October, 1980. For implementing the programme, DRDA was set up at district level. IRDP envisages the integration of methodology and approach of both beneficiary oriented programmes as well as area development efforts for the purpose of poverty alleviation as well as increasing productivity.

The goal of poverty alleviation was achieved by having two main instruments

(i) a set of self employment schemes for the poor *i.e.*, IRDP and its two sub-programmes of TRYSEM and DWCRA.

(ii) Wage employment programmes like NREP (1980) and RLEGP (1983). These have now been merged into Jawahar Rozgar Yojana (JRY).

While the programs under set (i) aim at giving the poor family an income generating asset, the latter set (ii) provides direct income to the poor through wage employment.

Objectives of IRDP

The main objectives are creating assets, employment, increased income, removal of poverty and minimizing inequality. The guide lines given to achieve the objectives include:

(i) To provide gainful employment and increase the purchasing power of rural poor.

(ii) The job opportunities must be provided through the application of science and technology in making optimum use of existing local resources.

(iii) The programme must be simple enough to operate and be economically viable to ensure quick self-reliance of its beneficiaries.

For achieving its objective of rural poverty alleviation, IRDP aimed at the provision of assets to the identified poor in the form of Government subsidy and bank credit on reasonable rate of interest.

Operation

For managing the programme a corporate governmental agency name as District Rural Development (DRDA) was set up at district level. The DRDA is guided and directed, and supported by a governing council headed by the Project Officer of DRDA. The District Collector, Heads of District Officers, legislators, Panchayat Union Chairmen and some other non-officials from the member.

The existing development block was strengthened to enable it to bear the additional responsibility of implementing DRDA directed and supported programmes. IRDP is a centrally sponsored scheme implemented by DRDA of the states. The scheme is funded on 50:50 bases by the centre and the states.

The integration involves several categories of

(i) Spatial integration (integration between areas)

(ii) Sectorial integration (integration between agriculture, off-farm activities, industries, etc.)

(iii) Integration in economic and social development.

(iv) Integration of total area and target group approach.

(v) Integration of human and other resources.

(vi) Integration of income generating schemes.

(vii) Integration of credit with technical services.

Thus IRDP involves integration both in its means and ends.

Short comings

Since 1985 the Government has promoted concurrent evaluation of the implementation and impact of the programme. These evaluations had indicated that there has been some increase in income, but only a minority of them could cross the newly determined poverty line of 6400 rupees. Various evaluations indicated the followings:

- The resources provided were inadequate in majority of the cases;
- There were large number of over dues due to one reason or the other;
- Poor quality of assets provided to the beneficiaries in many cases;
- The follow up of the beneficiaries was inadequate;
- Delay in provision of assets and preliminary high costs incurred by beneficiaries;
- Lacking of training facilities to farmers;
- Delay in releasing subsidy;
- Bribes taken by various functionaries concerned;
- Very inadequate supporting facilities or services;
- Non-availability of loans for making capital; and
- Lack of guidance about insurance cover.

Selection of beneficiaries the 'very poor' are overlooked, by and large.

9.2.3 National Health Rural Mission

About NHM

The Union Cabinet vide its decision dated 1st May 2013 has approved the launch of National Urban Health Mission (NUHM) as a Sub-mission of an over-arching National Health Mission (NHM), with National Rural Health Mission (NRHM) being the other Sub-mission of National Health Mission.

NHM has six financing components:

(i) NRHM-RCH Flexipool,

(ii) NUHM Flexipool,

(iii) Flexible pool for Communicable disease,

(iv) Flexible pool for Noncommunicable disease including Injury and Trauma,

(v) Infrastructure Maintenance and

(vi) Family Welfare Central Sector component.

ORGANOGRAM FOR NRHM IMPLEMENTATION AT STATE LEVEL

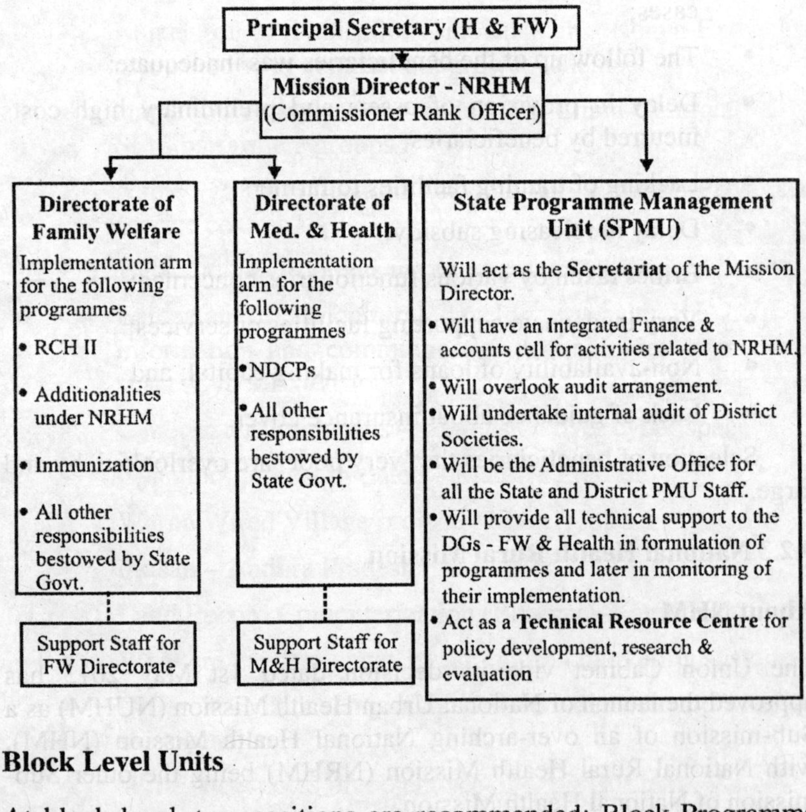

Block Level Units

At block level, two positions are recommended: Block Programme Manager; and Accounts Assistant

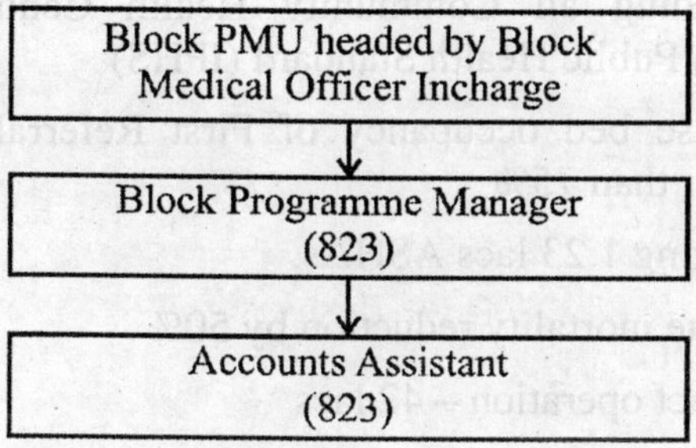

Mission

Improved health status and quality of life of rural population with unequivocal and explicit emphasis on sustainable development measure.

Objectives

Basic objectives for implementation of NRHM are:

- To reduce infant mortality rate and maternal mortality rate
- To ensure population stabilization
- To prevent and control of communicable and non-communicable diseases

Key Performance Indicators (to be achieved by 2012)

- To reduce Maternal Mortality Rate (MMR) to 258/lacs live births
- To reduce Infant Mortality Rate (IMR) to 36/1000 births
- To reduce Total Fertility Rate (TFR) to 2.8
- Malaria mortality reduction rate by 60%
- Kala-Azar mortality reduction rate by 100%
- Filaria/Microfilaria reduction by 80%
- Leprosy Prevalence Rate – Less than 1 per 10000
- Tuberculosis DOTS series – 85% cure rate and 70% detection of new sputum smear positive cases

- Upgrading all Community Health Centres (CHCs) to Indian Public Health Standard (IPHS)
- Increase bed occupancy of First Referral Units (FRUs) greater than 75%
- Engaging 1.23 lacs ASHAs.
- Dengue mortality reduction by 50%
- Cataract operation – 42 lacs
- Under National Blindness Control Programme (NBCP), objective is to reduce the prevalence rate from 1% to 0.5 by year 2012.
- To bring down Total Goitre Rate (TGR) to less than 10%.
- To ensure that more than 90% households consume iodized salt.
- To ensure availability of AYUSH by ensuring that at each block primary health centre (PHC), at least 2 Medical Officers (MOs), one of them AYUSH practitioner, are available all the time.
- Safe drinking water and sanitation facilities to greater than 60% of villages.
- Reduction of malnourished children by half of present level.
- Grade AYUSH (Aurvedic Yoga Unani Siddh and Homopath) for promotion of healthy life style

9.2.4 MNREGA

The Mahatma Gandhi National Rural Employment Guarantee Act (MGNREGA), 2005, was notified on 7 September 2005.

Mandate and Objectives

The mandate of the Act is to provide 100 days of guaranteed wage employment in a financial year (FY) to every rural household whose adult members volunteer to do unskilled manual work.

The objectives of the programme include

- Ensuring social protection for the most vulnerable people living in rural India through providing employment opportunities
- Ensuring livelihood security for the poor through creation of durable assets, improved water security, soil conservation and higher land productivity
- Strengthening drought-proofing and flood management in rural India,
- Aiding in the empowerment of the marginalized communities, especially women, Scheduled Castes (SCs) and Scheduled Tribes (STs), through the processes of a rights-based legislation,
- Strengthening decentralized, participatory planning through convergence of various anti-poverty and livelihoods initiatives
- Deepening democracy at the grass-roots by strengthening the Panchayati Raj Institutions (PRIs)
- Effecting greater transparency and accountability in governance

Salient Features of the Act

Registration: Adult members of a rural household willing to do unskilled manual work, may apply for registration either in writing, or orally to the local Gram Panchayat (GP). The unit for registration is a household. Under the Act, each household is entitled to a 100 days of employment every year.

Job Card: After due verification of place of residence and age of the member/s (only adult members are eligible for employment), the registered household is issued a Job Card (JC). Job Card forms the basis of identification for demanding employment. A JC is to be issued within 15 days of registration. Each JC has a unique identification number. The demand for employment in the GP, or at block level has to be made against the JC number. Job Cards are also supposed to be updated with days of work and payment made to the beneficiary as and when the work is undertaken.

Application for Work: A written application seeking work is to be made to the GP or Block Office, stating the time and duration for which work is sought. The GP will issue a dated receipt of the written application for employment, against which the guarantee of providing employment within 15 days operates.

Unemployment allowance: In case employment is not provided within 15 days, the state (as per the Act) will pay an unemployment allowance to the beneficiary.

Provision of Work

While allocating work, the below mentioned considerations are followed: Work is provided within 5 kilometres (kms) radius of the village. In case, work is provided beyond 5 kms, extra wages of 10 per cent are payable to meet additional transportation and living expenses. Priority is awarded to women, such that at least one-third of the beneficiaries under the Scheme are women. At least 50 per cent of works, in terms of cost, are to be executed by the GPs.

Wages

Wages are to be paid as per the State-wise Government of India (GoI) notified MGNREGA wages. Wages are also to be paid according to piece rate, as per the Schedule of Rates 4 (SoRs). Payment of wages has to be done on a weekly basis and not beyond a fortnight in any case. Payment of wages is mandatorily done through the individual/joint bank/post office beneficiary accounts.

Planning

Plans and decisions regarding the nature and choice of works to be undertaken in a FY along with the order in which each work is to be taken up, site selection, etc. are all to be made in open assemblies of the Gram Sabha6 (GS) and ratified by the GP. Works that are inserted at Block and District levels have to be approved and assigned a priority by the GS before administrative approval can be given.

Cost Sharing

The GoI bears the 100 per cent wage cost of unskilled manual labour and 75 per cent of the material cost, including the wages of skilled and semi-skilled workers

Worksite Management

To ensure that the workers are directly benefitted under the Scheme, the Act prohibits the use of contractors or machinery in execution of the works. To ensure that the spirit of the Act is not diluted and wage employment is the main focus, MGNREGA mandates that in the total cost of works undertaken in a GP, the wage expenditure to material expenditure ratio should be 60:40. Worksite facilities such as crèche, drinking water and shade have to be provided at all worksites.

Transparency and Accountability

Transparency and accountability in the programme is ensured through the following: Social audit, 7 to scrutinize all the records and works under the Scheme are to be conducted regularly by the GS. Grievance redressed mechanisms and rules have to be put in place for ensuring a responsive implementation process. All accounts and records relating to the Scheme should be available for public scrutiny.

Funding

The Central Government bears the costs on the following items:

- The entire cost of wages of unskilled manual workers.
- 75% of the cost of material, wages of skilled and semi skilled workers.
- Administrative expenses as may be determined by the Central Government, which will include inter alia, the salary and the allowances of the Programme Officer and his supporting staff, work site facilities.
- Expenses of the National Employment Guarantee Council.
- The State Government bears the costs on the following items:
- 25% of the cost of material, wages of skilled and semi skilled workers. Unemployment allowance payable in case the State Government cannot provide wage employment on time. Administrative expenses of the State Employment

Guarantee Council. Districts have dedicated accounts for NREGA funds. They have submitted their proposals based on clearly delineated guidelines so that funds may be distributed efficiently at each level, and adequate funds may be available to respond to demand. Under NREGA, fund releases are based on an appraisal of both financial and physical indicators of outcomes.

9.2.5 Rajiv Gandhi Scheme for empowerment of Adolescent girls/Boy's

Sponsored by: Central Government

Beneficiaries: Individual, Women

How to Avail: Information is available at ICDS Offices and Anganwadi Centers (AWCs).

Introduced: On 01/04/2011 Valid Upto 31/03/2015.

Eligibility criteria

The Scheme would cover adolescent girls in the age group of 11–18 years under all ICDS projects in selected 200 districts in all the States/UTs in the country. In order to give appropriate attention, the target group would be subdivided into two categories, viz. 11–15 and 15–18 years and interventions planned accordingly.

Rajiv Gandhi Scheme for Empowerment of Adolescent Girls (RGSEAG) – SABLA is a scheme to empower adolescent girls of 11–18 years by improving their nutritional and health status, upgrading various skills like home skills, life skills and vocational skills. The girls will also be equipped with information on health and family welfare, hygiene and guidance on existing public services. The scheme is being implemented in 200 districts across the country on a pilot basis while in the remaining districts Kishori Shakti Yojana (KSY), where operational, will continue as before. The scheme is being implemented using the platform of Integrated Child Development Scheme (ICDS) wherein Anganwadi Centres (AWC) would be the focal point for the delivery of services.

In the Indian context, adolescent girls are vulnerable to physical and educational neglect. Malnutrition is more common in India than in Sub-Saharan Africa with over 90 per cent of young girls reported as anemic. Academically, adolescent girls have a much higher rate of school dropout than boys. Such neglect becomes an impediment to the progress of young girls and women in society. The RGSEAG-SABLA program works to combat these issues by focusing on educational, nutritional and health requirements

GSEAG refers to Rajiv Gandhi Scheme for Empowerment of Adolescent Girls, which was launched in 2010. The scheme is being implemented in 200 districts across the country on pilot basis. This is a Centrally Sponsored Scheme for adolescent girls from 11 to 18 years also known as SABLA. It is being implemented using the platform of Integrated Child Development Services Scheme. Anganwadi Centres are the focal point for the delivery of the services. The scheme has two major components namely nutrition and non-nutrition component. Nutrition is being given in the form of Take Home Ration or Hot Cooked Meal for 11 to14 years out of school girls and 14 to18 years to All AGs, out of school and in school girls. In the Non-nutrition component, the out of school Adolescent Girls 11 to18 years are being provided IFA supplementation, Health check-up and Referral services, Nutrition and Health Education, Counseling and guidance on family welfare, Adolescent Reproductive Sexual Health, child care practices and Life Skill Education and accessing public services and 16 to 18 year old Age are also being given vocational training, Anganwadi centre AWC in the village is the focal point for the implementation of the scheme SABLA, where the school going girls and out of school girls would meet. At the AWC they receive life skill education.

Description

An integrated package of services is to be provided to Adolescent Girls that would be as follows:

i. Nutrition provision

ii. Iron and Folic Acid (IFA) supplementation

iii. Health check-up and Referral services

iv. Nutrition and Health Education (NHE)

v. Counseling/Guidance on family welfare, ARSH, child care practices and home management

vi. Life Skill Education and accessing public services

vii. Vocational training for girls aged 16 and above under National Skill Development Program (NSDP)

Funding Pattern

Centrally sponsored scheme, implemented through the State Governments/UTs with 100 per cent financial assistance from the Central Government for all inputs, except nutrition provision for which Government of India will share up to the extent of 50 per cent of the financial norms or the actual expenditure incurred, whichever is less.

The objectives of the Saksham scheme would be to facilitate, educate and empower ABs so as to enable them to become self-reliant, gender-sensitive and aware citizens. The scheme will have the following objectives –

i. To make the Adolescent Boys gender sensitive.

ii. To create sensitized Ahimsa Messengers to address Violence against Women.

iii. To enable ABs for self-development and empowerment.

iv. To address the health needs *i.e.* the physical, mental and emotional health of ABs.

v. To promote awareness about health, hygiene, nutrition and Adolescent Reproductive and Sexual Health (ARSH) and family and child care.

vi. To provide appropriate information and vocational skills for ABs above 16 years through National Skill Development Program (NSDP) for future work-participation.

vii. To provide necessary life skill education and to provide information/guidance about existing public services.

viii. To channelize the energies of ABs for nation building.

Extent of the problem

One in every three malnourished children in the world lives in India. Anemia affects 90 per cent of adolescent girls in India.

- According to the Ministry of Human Resource Development gross enrollment sex ratio in primary

- A school is almost 1 but in secondary education it drops down to 0.79, 80 per cent of women are engaged in informal sectors, primarily agriculture.

9.2.6 Gramin Bhandaran Yojana (GBY)

Introduced – On 31/03/2007 Valid Upto 31/03/2012

Sponsored by – Ministry of Agriculture, GoI

Funding Pattern – The funds for this scheme are contributed by National Bank for Agricultural and Rural Development, National Cooperative Development Corporation and by cooperative banks and cooperative societies

Ministry/Department – Department of Agriculture Camp; Cooperation

How to Avail – An interested promoter will submit a copy of the project proposal for term loan and subsidy to the respective banks on an application form as prescribed by the concerned bank along with project report and other documents for appraisal and sanction of loan.

The Gramin Bhandaran Yojana, announced in 2001, is a capital investment subsidy scheme for renovation and construction of rural godowns. Godowns should, however, be constructed outside municipal corporation limits. It is important to note that rural godowns (promoted by Ministry of Food Processing) which are located in food parks can also avail of Gramin Bhandaran Yojana.

What's the purpose of Gramin Bhandaran Yojana?

The Gramin Bhandaran Yojana envisages a robust network of non-urban godowns to help build adequate scientific storage for farm produce in rural areas. Owing to enhanced storage capacity, farmers in villages are less likely to opt for distress sale of their harvest courtesy marketing credit and pledge financing available under the scheme. The purpose of the scheme is to cater to better storage facilities of farm produce, agricultural inputs, and processed farm produce besides standardization of farm produce to boost its demand in the market and pave way for a system of warehouse receipts on a national level. The scheme is aimed at arresting the decreasing investment ratio in the agriculture sector in the country by extending subsidy to cooperative and private sectors.

The following are eligible to avail of the Gramin Bhandaran Yojana

- Marketing boards
- Agro-processing co-operative societies
- Quality testing laboratories
- Agro-processing corporations
- Partnership firms
- Agro-industrial corporations
- Companies
- Agricultural Produce Marketing Committees
- Farmers
- Proprietary firms
- Co-operatives
- Non Governmental Organisations
- Corporations
- Farmers' groups
- Self-help groups

Features of Gramin Bhandaran Yojana

Size of godown: Owner has to decide the capacity of his or her godown. Subsidy for the scheme is provided for a minimum potential of 100 loads and maximum of 30,000 loads. A smaller rural godown with 50 ton capacity is also qualified under the scheme on the basis of topography or viability analysis/requirement of a region. Also, godowns located in hilly areas with 25 ton capacity also qualify for the scheme in specific cases.

Pledge Loan: Upon hypothecation of farm produce, farmers can avail of the pledge loan facility. The interest rates, loan amount and pledge period are based on the RBI/NABARD guidelines and banking practices of various financial institutions.

Conditions: Godowns should be robust in terms of their structure for storage of agricultural produce. Entrepreneurs may have to get a license under the State Warehousing Act. Godowns which have capacity to store around 1,000 tonnes must be accredited by the Central Warehousing Corporation (CWC).

Location: Godowns must be constructed outside the limits of municipal corporation. Also, any godowns (promoted by Ministry of Food Processing Industries) located at food parks are eligible as well.

Credit-linked assistance

Subsidy linked to institutional credit is available for projects financed by the following:

- Urban cooperative banks
- Regional rural banks
- Commercial banks
- North Eastern Development Finance Corporation (NEDFI)
- State Co-operative Agricultural and Rural Development Bank
- State cooperative banks
- Agricultural development finance companies

Credit facility will be extended for construction costs of boundary wall, grading, platform, packaging, internal road, internal drainage system in addition to quality certification and warehousing facilities

Capital Cost

Godowns (1000 tonnes): Depends on the actual cost or cost of a project appraised by a bank or Rs.3500 per ton, whichever is lower. Godowns above 1000 tonnes capacity: Depends on the actual cost or cost of a project appraised by a bank or Rs.1500, whichever is lower.

Subsidy

- 25 per cent of a project's cost for farmers and cooperatives subject ceiling on Rs. 2.25 crores (subsidy)

- 33.33 per cent of a project's cost for SC/ST entrepreneurs and projects in hilly areas subject to ceiling on Rs. 3 crores (subsidy)
- 15 per cent of a project's capital cost for individuals, corporations, subject to ceiling on Rs.1.35 crores (subsidy)
- 25 per cent of a project's cost for godowns' renovation aided by NCDC

Security: Mortgage of land and godown

9.2.7 Pradhan mantri Adarsh Gram Yojana (PMAGY)

Pradhan Mantri Adarsh Gram Yojana (PMAGY), a Government of India initiative for the empowerment of deprived sections, aims to achieve integrated development of selected villages through convergent implementation of all relevant Central and State schemes. The scheme was launched in March, 2010 on a pilot basis for the integrated development of 1000 villages each with more than 50 per cent SC population. Vision of an "Adarsh gram" (Model village).

A Model village is one which has adequate physical and institutional infrastructure, in which minimum needs of all sections of the society are fully met; they live in harmony with each other, as also with the environment, and a village which is progressive and dynamic. These villages should be covered with all the facilities necessary for dignified living, creating thereby an environment in which all its residents are enabled to utilise their potential to the fullest. These villages should, inter alia, satisfy the following norms:

Physical Infrastructure

- Should be connected to the nearest major road by an all-weather road. Likewise, in case of a multi-hamlet village, all hamlets should be connected with each other by an all-weather road.
- Access for all too, safe drinking water on a sustainable basis.
- All houses should have electricity.
- The village should have slush-free internal roads, and adequate street lighting.

- Village should have adequate communication facilities, such as post-office, telephones, and, if possible, internet, and Bharat Nirman Common Service Centre (being established by the Department of Information Technology).
- Availability of adequate banking facilities through regular (brick and mortar) branches in the village or in close proximity, and through Business Correspondent/Business Facilitator Model.
- All residents should have adequate housing, and there should be no homeless family.

Sanitation and Environment

- The village should have a high degree of sanitation. It should be free from dry latrines, and open defecation, and should have sanitary toilets, drains and an efficient waste disposal system. It should, as far as possible, fulfill "Nirmal Gram Puraskar" norms.
- The Village should take care of its environment through (1) Planting trees, (2) water harvestation and maintenance of water bodies (3) use of renewable sources of energy, such as biogas, solar energy, wind energy, (4) use of smokeless chulhas, etc.

Social Infrastructure, Human Development and Social Harmony

- Should have an Anganwadi centre and schools of appropriate levels.
- The village should have adequate and attractive building for its anganwadi, school, health centre, Panchayat, and community hall. The village should have adequate facility for sports and other physical activities.
- All children in the age-group of 3–6 should be enrolled in, and regularly attend the Anganwadi. Likewise, all children in the 6–14 age group should be enrolled in and regularly attend school.
- All adults should be at least functionally literate, and should have access to facilities for continuing education.

- Access for all primary health care and Reproductive Child Health (RCH) facilities, with proper pre-natal and ante-natal care for mothers.
- 100 per cent institutional deliveries, full immunization of children, and observance of the small family norm.
- The village as a community should take special care of its women, children (especially girls), senior citizens, and persons with disabilities.
- There should be no public consumption of liquor or any other intoxicating substances, and their use in general should be discouraged.
- The village should have an active Gram Sabha/Gram Panchayat, women's/Swarozgaris' Self-help Group, youth club and Mahila Mandal.
- There should be no caste-based discrimination, complete eradication of untouchability, and a due sense of security and dignity among the weaker sections.
- Residents of the village should be aware of and exercise their constitutional and legal rights. Likewise, they should also be aware of and discharge their fundamental and civic duties.

Livelihood

- All youth and adults of the village should have adequate employment and means of livelihood, and there should be adequate arrangements for development of skills among them, so that as many of them as possible are in skilled employment.
- Progressive and efficient practices, based on new technology, should be used in all the economic activities pursued in the village, especially in agriculture, animal husbandry, fisheries etc.
- Should have adequate access to remunerative prices for agricultural and other produce of the village.

Objective

To ensure integrated development of the selected villages into "Model villages" so that, inter alia,

1. They have all requisite physical and social infrastructures for their socio-economic development, and satisfy the norms mentioned in the vision of an *adarsh gram* to the maximum possible extent.
2. Disparity between SC and non-SC population in terms of common socio-economic indicators (*e.g.*, literacy rate, completion rate of elementary education, IMR/MMR, ownership of productive assets, etc.) is eliminated, the indicators are raised to at least the level of the national average, and

- All BPL families, especially those belonging to SCs, have food and livelihood security, and are enabled to cross the poverty line and earn an adequate livelihood
- All children complete at least eight years of education, and
- Incidence of malnutrition, especially among children and women, is eliminated
- Untouchability, discrimination, segregation, and atrocities against SCs are eliminated, as are other social evils like discrimination against girls/women, alcoholism and substance (drugs) abuse, etc., and all sections of society are able to live with dignity and equality, and in harmony with others.

Area covered

Presently the scheme is being implemented in five States of the country *viz.* Assam, Bihar, Uttar Pradesh, Rajasthan and Tamil Nadu.

Implementation

The scheme is being implemented by the Ministry of Social Justice and Empowerment, Government of India. To ensure focused attention, the selected States are to select villages from same, or 2–3 contiguous districts and give preference to the more backward districts. The development of the villages has to be ensured through:

- Convergent implementation of existing Schemes of the Centre and State Governments, and
- The works which could not be covered under the above, to be taken up through provision of "gap-filling" funds for which central assistance is provided @ Rs. 10 lakh per village revised to Rs. 20 lakhs per village at an average w.e.f. Sept. 2011, with State Government also expected to make a suitable, preferably matching, contribution.

Based on the experience gained in the pilot phase, the scheme will be considered for implementation on a larger scale.

As many as 1,500 Scheduled Caste majority villages across 10 states, including Uttar Pradesh and Punjab in the north and Assam, West Bengal and Jharkhand in eastern India, will be now benefitted from the Pradhan Mantri Adarsh Gram Yojna (PMAGY).

Launched in 2009–10, the Pradhan Mantri Adarsh Gram Yojana (PMAGY) is a centrally sponsored scheme which is implemented by the Ministry of Social Justice and Empowerment for integrated development of Scheduled Caste majority villages having Scheduled Caste population of more than 50 per cent.

In the northern region, Uttar Pradesh, having the maximum number of SC majority villages, "did not concur" to accept the scheme in the first phase and so Himachal Pradesh was selected for the implementation of the scheme, an official source said her.

9.2.8 Pradhan Mantri Kaushal Vikas Yojana

Ministry – Skill Development and Entrepreneurship

Launched – 16 July, 2015, New Delhi

"Skilling is building a better India. If we have to move India towards development then skill development should be our mission"

Objectives – The objective of this scheme is to encourage and promote skill development for the youth throughout the country by aligning itself with the Common Norms guidelines. The scheme also needs to be aligned to complement all other missions of the Government, such as Make in India, Digital India, Swachh Bharat, and Smart Cities. Specifically, the Scheme aims to:

- Enable and mobilize a large number of youth to take up industry designed quality skill training, become employable and earn their livelihood. The increase in productivity of the existing workforce and align skill training, able to achieve the actual needs of the country.
- Encourage standardization of the Certification process and put in place the foundation for creating a registry of skills.
- Benefit 10 million youth over the period of four years (2016–2020)

KEY COMPONENTS OF THE SCHEME

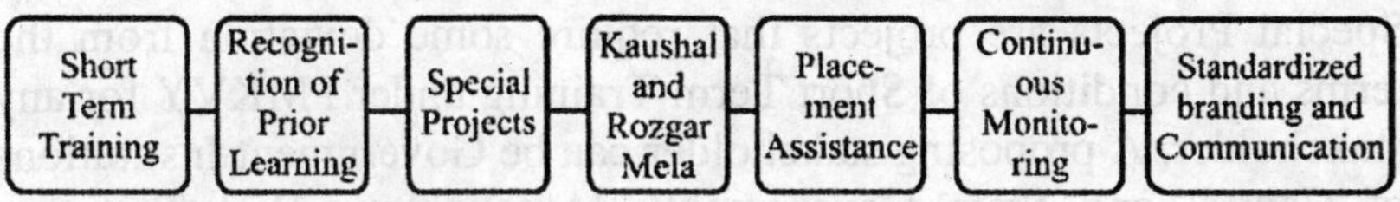

Key Components of the Scheme

1. Short Term Training

The Short Term Training imparted at PMKVY Training Centres (TCs) is expected to benefit candidates of Indian nationality who are either school/college dropouts or unemployed. Apart from providing training according to the National Skills Qualification Framework (NSQF), TCs shall also impart training in Soft Skills, Entrepreneurship, Financial and Digital Literacy. Duration of the training varies per job role, ranging between 150 and 300 hours. Upon successful completion of their assessment, candidates shall be provided placement assistance by Training Partners (TPs). Under PMKVY, the entire training and assessment fees are paid by the Government. Payouts shall be provided to the TPs in alignment with the common norms. Trainings imparted under the Short Term Training component of the Scheme shall be NSQF Level 5 and below.

2. Recognition of Prior Learning

Individuals with prior learning experience or skills shall be assessed and certified under the Recognition of Prior Learning (RPL) component of the Scheme. RPL aims to align the competencies of the unregulated workforce of the country to the NSQF. Project Implementing Agencies (PIAs), such as Sector Skill Councils (SSCs) or any other agencies designated by MSDE/NSDC, shall be incentivized to implement RPL projects in any of the three Project Types (RPL Camps, RPL at Employers Premises and RPL centres). To address knowledge gaps, PIAs may offer Bridge Courses to RPL candidates.

3. Special Projects

The Special Projects component of PMKVY envisages the creation of a platform that will facilitate trainings in special areas and/or premises of Government bodies, Corporate or Industry bodies, and trainings in special job roles not defined under the available Qualification Packs (QPs)/National Occupational Standards (NOSs). Special Projects are projects that require some deviation from the terms and conditions of Short Term Training under PMKVY for any stakeholder. A proposing stakeholder can be Government Institutions of Central and State Government(s)/Autonomous Body/Statutory Body or any other equivalent body or corporates who desire to provide training to candidates.

4. Kaushal and Rozgar Mela

Social and community mobilization is extremely critical for the success of PMKVY. Active participation of the community ensures transparency and accountability, and helps in leveraging the cumulative knowledge of the community for better functioning. In line with this, PMKVY assigns special importance to the involvement of the target beneficiaries through a defined mobilization process. TPs shall conduct Kaushal and Rozgar Melas every six months with press/media coverage; they are also required to participate actively in National Career Service Melas and on-ground activities.

5. Placement Guidelines

PMKVY envisages linking the aptitude, aspiration, and knowledge of the skilled workforce it creates with employment opportunities and demands in the market. Every effort thereby needs to be made by the PMKVY TCs to provide placement opportunities to candidates, trained and certified under the Scheme. TPs shall also provide support to entrepreneurship development.

6. Monitoring Guidelines

To ensure that high standards of quality are maintained by PMKVY TCs, NSDC and empanelled Inspection Agencies shall use various methodologies, such as self-audit reporting, call validations, surprise visits, and monitoring through the Skills Development Management System (SDMS). These methodologies shall be enhanced with the engagement of latest technologies.

The scheme will be implemented through the National Skill Development Corporation (NSDC).

- Approved for another four years (2016–2020) to benefit 10 million youth.
- Allocated Budget 12,000 Crores.

Important questions

- Writes the objectives of rural development
- Depict the problems in Rural Development
- Explain notes on Integrated Child Development Scheme
- When was launched IRDP-1978
- Differences CDP and IRDP
- Explain the MNREGA
- When was existence MNREGA act was notified on - 7 September, 2005
- When was introduced RGSEAG – 4, April, 2011
- SABLA scheme to empower adolescent girls age between - 11–18 years
- SABLA scheme is implemented in 200 districts across India
- SABLA scheme 100% funded by Central Government
- Explain the objectives of RGSEAG
- Write the short notes of Gramin Bhandaran Yojana
- Write down the features of GBY
- Which scheme related to model village, explain it

- Write down the objectives of PMAGY
- Pradhan Mantri Kaushal Vikas Yojana Scheme launched by and in Minisrty of Skill development and entrepreneurship in 16 July, 2015
- Write the components of the PMKVY scheme

■■■■

10

Community Development

10.1 MEANING, DEFINITION

Community: It is a group of people who lives in a geographical area and have an interest in each other for the purpose of making a living.

Development: It connotes growth or maturation. It implies gradual and sequential phases of change. It refers to the upward or increasing differentiation.

Community Development: It is a movement designed to promote better living for the community with the active participation and/or the initiative at the community.

10.2 CONCEPT AND PRINCIPLES OF COMMUNITY DEVELOPMENT

Community Development Project (CDP-1952): India was a prosperous country before the British Rule. It was known as a golden bird throughout the world. Even the historians state that there were the rivers of milk flowing in this country. At that time the main reason for the prosperity was the rural based economy. All the villages were self-sufficient. In villages there were cottage industries, small industries, animals were domesticated and agriculture was done. The basic needs of that area were fulfilled by these occupations. All the villagers were self-sufficient and due to the joint family system the members of family shared their work, so the economic condition of the family was good and strong. When the British understood this cause of prosperity of rural people they concluded that without uprooting the rural economy, they cannot establish English trade and industry and they cannot rule over this country. So, they uprooted many Indian rural industries and introduced industrialization and use of machines in the country with the introduction of technology in the country, the rural individual based economy was destroyed. The British government seized over the small industries *i.e.,* cotton, soap, pin and manure processing. So as the social and economic system of the country was ruined, the joint family began to separate. This tragic condition continued till about 200–250 years in the country.

In 1857, the freedom fighters initiated a long struggle against the rulers. In the contemporary period the leaders and social reformers urged the people to use things made in our country, but their influence on the people was not powerful enough. Lokmanya, Bal Gangadhar Tilak, Mahatma Gandhi, Rabindra Nath Tagore, and Dr. Keshava Rai Hedgewar etc. great personalities began this Swadeshi Movement and thus the society became enlightened.

In spite of suppressive nature of the English people, many social workers introduced various movements for integrated rural development. Its main objective was that all the rural people of the village co-operate for the rural development programme. The feeling of Community must be developed among the rural people. In the chain of these programmes there were *"Seva Gram"* a project of Mahatma Gandhi in 1920, "Shree Niketan Project" by Rabindra Nath Tagore in 1920, "Marthandam Project" of Dr. Spencer Hatch in 1928, "Baroda Village Reconstruction Project" of Dr. B.T. Krishnamachari in 1932, Indian Village Services of Dr. W.H. Wisher in 1945, and Firka Development Project of the government of Madras in 1946 etc. projects. All these social worker's basic objectives were to develop the integrated village. After attaining independence, the Government of India has undertaken many efforts to make the above concept more effective and among them Etawah Pilot project proved very effective. From this Project the Government of India achieved many encouraging results. After getting the results from the projects, the government introduced a new scheme on 2nd October in 1952 and this project was called *"Community Development Project"*. According to the scientists there should be established such an organisation in the country whose main objective should be the all-round development of the villagers and for achieving this aim, all the country men should co-operate. In the rural field, the resources of human development should be developed. In this direction, everyone knew about the successful results of Mahatma Gandhi's project, Ravindra Nath Tagore's project and Etawah Pilot Project and so considering all these points, this project was introduced. In this context, Pt. Jawahar Lal Nehru had stated in the fourth All India Conference of Development Commissioners that *"These community schemes are not a replica or a copy of something from abroad, although we have learnt much from other countries. They are essentially an Indian growth suited to India's conditions"*. The main philosophy of this project was Co-operation and Self Help. This concept will make the country a new India. For the reconstruction of India, the Planning Commission, considering the basic concepts for the community development, stated (1) Community Development is a method, (2) National Extension Service is an agency, and (3) The goal is to change the social and economic condition of village. Thus community development project is an agency to change the social and economic conditions of the villages.

Objectives

(1) To increase the agricultural production. (2) Community and integrated development. (3) The extension of the new scientific knowledge. (4) Development of small and medium irrigation projects. (5) Development of co-operative organisations. (6) Construction of roads. (7) To increase the adult education and primary education. (8) Facility for entertainment. (9) Development and construction of primary health care centre and the public health service. (10) To inspire the youth for the development programme.

10.3. PRINCIPLES OF COMMUNITY DEVELOPMENT

Canadians Michael and Julie Bopp described the principles of community development as follow:

1. **Harnessing Community Tensions:** For changes to occur there must be a tension in people's mind between the way things are and the way people want them to be. There must be a desire for change and a desire to work together to create the change.

2. **Facilitating Consultation about Community Realities and Needs:** As noted in detail below, for community development to work successfully, there has to be awareness about community needs as decided upon by the community themselves, not as decided upon by outsides.
3. **Maintaining Unity and Healthy Human Relations:** It is important to build respectful healthy relationship between members of the group and to work on managing conflicts as they occur so as to maintain a collective vision.
4. **Developing a Common Vision of Sustainable Future:** If community development is to work, it must be sustainable and the process must include everyone. Stories abound of development processes failing because the resources or the process were not sustainable or well devised or were done without consultation. We have all seen pictures of computers lying unused because the hardware or software broke or just because people in the community didn't actually see need for computers.
5. **Supporting Core Group Development:** It is essential to encourage and support your group. Get to know the people in your group. What are their interests and skills? How much time can they dedicate to the group's activities? Everyone needs to be clear about their capacities and the limits to their time and energy that may exist. Encourage and build up a strong support base so that the group is not dependent upon only one or two people. Make sure the group is as representative of the community as it can be, that means including and listening to the needs of women and children as well.
6. **Personal Revitalization and Healing**: Good community development processes provide an opportunity for people to become empowered, energized and to undo some of the harms of the past.
7. **Facilitating Learning:** The entire development process is a learning experience. People can become aware of their environment, their social and political reality and can learn how to make the steps necessary to change and improve their situation.

8. **Building Effective Organisations:** Formalizing the initial volunteer group is crucial part of the development process. Creating and implementing formal organisation structures create legitimacy and respect for the organisations and lets the community know that you are serious about your endeavours. It helps keep people unified and accountable and is part of the process of developing and implementing plans and activities.

9. **Networking with Resources and Allies:** Creating the networks between likeminded individuals who then form a group, who then network with other groups in order to learn and share skills as well as resources and knowledge, is a crucial part of the community development process. The most effective groups are those that are well networked and work on co-operation with others in their area. Find out whether your area has a friendship school or rotary club or other group that is already working.

10. **Programme Development:** It is important to have a clear programme. An adhoc approach to community developm-ent processes will not work. A programme must be devel-oped in conjunction with the community and be about community needs and priorities. They must adhere to real-istic timelines and have inbuilt monitoring and evaluation processes so that changes can be tracked. The progarmme acts as a map to signal where it is the group going.

11. **Reflection on the Process (Monitoring and Evaluation):** Monitoring and Evaluation is very important as it allows the group to track progress, identify problems, and improve upon community development process.

12. **Protecting the Process:** Community development proce-sses are dependent upon a rant of other processes; the good will of those involved, the continuation of funds, a secure political environment, a lack of violence. Once any of these dynamics is disrupted, then negative change and conflict may occur. It is important to have risk management strate-gies built into your process less well developed processes might have fallen victim to the conflict and insecurity.

10.4. PHYSIOLOGY OF COMMUNITY DEVELOPMENT

Basic Philosophy: Basic philosophy of community development programme was in the following way: (1) Individual development. (2) Development of family. (3) Awareness for the responsibilities and self-motivation among the people. (4) Community development. (5) Development of Cooperativeness among the people. (6) To create the confidence towards science and technology. (7) Development of rural leadership. (8) Development of rural institution. (9) Development of other resources for community development. (10) The social, cultural and economic development of rural society.

Important questions

- Define the community
- Define Community Development
- Explain concept and Principles of Community development
- What are the objectives of CDP
- When was launched CDP - 2, Oct., 1952
- CDB was started in - 1954

■■■■

11

Panchayati Raj System

11.1 DEFINITION, MEANING

The realization that peoples' participation is crucial for successful implementation of programmes like CD and NES was brought to sharp focus through the report of the team for the Study of Community Projects and National Extension Service by Balwant Rai Mehta (1957). The Committee observed that one of the least successful aspects of CD and NES work is its attempt to evoke popular initiative and recommended democratic Decentralisation.

Democratic Decentralisation: The word *"Democracy"* is derived from the Greek roots *'Demos'* meaning 'the people' and 'Kratos' meaning 'Authority'. In democracy all authority originates from the people. It is governance of the people, by the people, for the people. Rule by majority is, no doubt, an important feature of this system of governance; but the more important ingredient is rule by consultation – consultation between the people's representatives on one hand and consultation with the officers on the other. This is popularly known as *"Panchayati Raj"* in India.

Decentralisation: Means devolution of central authority among local units close to the areas served. Where authority devolves by this process on people's institution, it is 'democratic decentralization'.

The State of Madras tried this as a pilot project as early as 1957. Based on the success in this State it was Rajasthan which became the pioneer to bring the whole state under democratic Decentralisation on Oct. 2, 1959.

Meaning - Panchayat Raj

The concept "democratic Decentralisation" was not easily understood by the people. Therefore, as decided by the then Prime Minister the three-tier administration was introduced in the name of "Panchayat Raj". Panchayat Raj means the system of Government. Horizontally it is a network of village Panchayats. Vertically, it is an organic growth of Panchayat rising up to national level.

Panchayat Raj was easily accepted because it meant administration by mutual consultation, consent and consensus. It fitted closely into the ancient cultural patterns in India.

Need for Democratic Decentralisation

India is a sovereign democratic republic. Till recently, our administration was described as 'democracy at the top and dictatorship at the bottom'. It was 'an inverted pyramid'. It was realized that, so long as the people who constitute the roots of the democratic system were kept apart from the administration at levels close to them, it would be impossible, to secure their support or enthusiasm for the Programme. The study team on CD Programmes headed by Balwant Rai Mehta recommended "Democratic Decentralisation".

Therefore *'Democratic Decentralisation'* ensures the direct participation of people at the grass root level. In 1956, the National Development Council appointed a committee under Balwant Rai Mehta, which submitted its report in 1957 in which it recommended the following.

Three tier structure consisting of Zilla Parishad at the District Level, Panchayat Samiti at the Block Level and Gram Panchayat at the Village Level.

The three tier system of Panchayat Raj was 1st adopted by Rajasthan (Nagaur District) on Oct 2, 1959. This was followed by Andhra Pradesh, Bihar Gujarat, Himachal Pradesh, Maharashtra, Punjab, Tamil Nadu, Uttar Pradesh and West Bengal.

11.2 THREE TIERS OF PANCHAYAT RAJ

The Gram Panchayat: The first formal democratic institution under the directive principle in the Indian constitution is the Gram Panchayat or Panchayat. It is the primary unit of local self government. Panchayat is a cabinet of the village elders, directly elected by the adult citizens of the village.

Gram Panchayats are constituted considering their income, population and area. The income varies from mere Rs. 500/- per annum to more than Rs. 2.00 lakhs. The population varies from 500 to 25,000. The Panchayat membership varies from 5 to 17. There is provision for reservation of seats for women and SC and STs. The Panchayat has tenure of five years and is directly elected. The meeting is to be convened at least once in six months. *The main functions of Panchayats are:*

(i) Representative functions, where the main role is to voice and represent the opinion;

(ii) Regulatory and administrative functions, which consist of regulating the conduct of individuals and institutions and also collection of taxes;

(iii) Service or developmental function, such as promotion of education, health, agriculture, etc.

The Panchayat Samiti or Panchayat Union: This is the second tier of the administration at Block level. Reservation and cooperation are given for women, SCs and STs and persons with experience in administration and public life. Block Development Officer is appointed by the Government. He functions as the leader of the Block.

Functions of the Block

(1) It has to instill among people within its jurisdiction a spirit of self-help and initiative and work for raising the standard of living;

(2) It has to support for the implementation of development programmes;

(3) It has the welfare and development activities in the fields of agriculture, animal husbandry, health, sanitation, elementary education, cottage industries and social.

(4) It has to use the village housing project funds and loans.

District Development Council (Zila Parishad)

This is the third tier of Panchayat Raj functioning at district level. The members are all Panchayat Union Chairman, District Collector, MLAs, MLCs and MPs of the district with right to vote, but not to hold office and women, SC, ST and persons interested in rural development are also given representation. District Collector leads the work with the help of District Development Officers.

Functions

It works as advisory body for blocks. It approves budget and plan of blocks. It allots funds to the blocks. Secondary education is the responsibility of this council. It should advise Government in all matters relating to rural development in the district. It has to review the results achieved under various items in all the blocks.

The functions of Panchayat Raj Institutions can be summarized as below:

Panchayat	Panchayat Union	District Development Council
Sanitation, conservation and water supply	Administration of blocks	Approval of Panchayat union budget
Construction and maintenance of roads, bridges, drains etc.	Execution of all programmes under CDP	Distribution of funds of Panchayat Union
Promotion of agricultural cooperative, cottage industries etc.	Approval of Panchayat budget	Coordinating the plans and supervising it
	Management of elementary schools.	Advise the Government for all the district rural development works
	Promotion of agriculture, cottage industries etc.	Secondary Education

Extension Organisation in Panchayat Raj Set-up

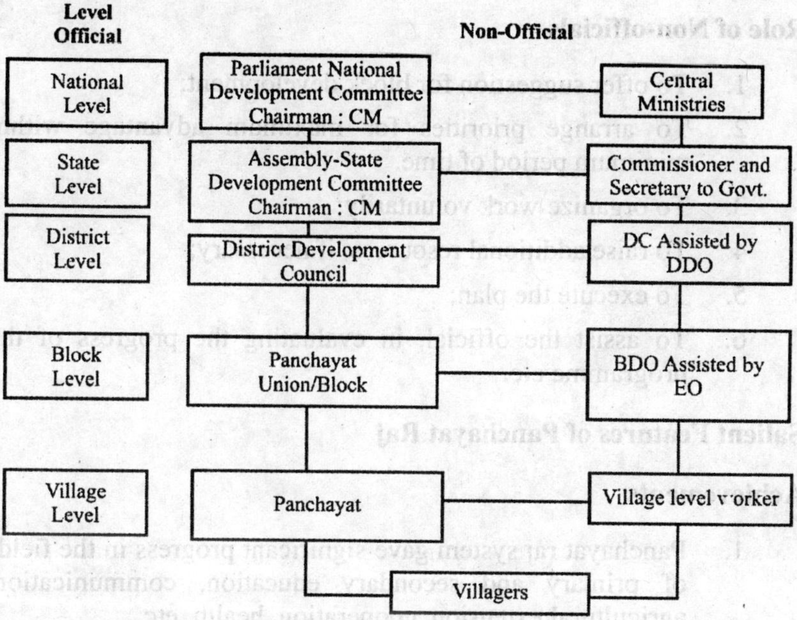

Role of Officials: No doubt that any programme is peoples' programme. Local leaders knew the people and know the local needs and potentialities. However, the officials are the most competent persons to judge what is good for people. So, their roles are:

1. To identify the felt needs of the people.
2. To fix priorities for peoples' needs, with their assistance;
3. To list out the local resources available;
4. To harness available and potential resources;
5. To encourage peoples' involvement in programmes;
6. To develop a community outlook among the people;
7. To educate the people to think in terms of planned development;
8. To mobilize required resources
9. To act as SMS at the time of programme development and implementation; and

10. To assist people in evaluating their progress.

Role of Non-officials

1. To offer suggestion for block development;
2. To arrange priorities for maximum advantage within minimum period of time.
3. To organize work voluntarily;
4. To raise additional resources, if necessary;
5. To execute the plan;
6. To assist the official, in evaluating the progress of the programme etc.

Salient Features of Panchayat Raj

Achievements

1. Panchayat raj system gave significant progress in the fields of primary and secondary education, communication, agricultural extension, cooperation, health, etc.
2. People could get drinking water. In some places people had protected water supply. Village streets, electric light provision, village sanitation, etc., had enough resources.
3. Rural awakening was brought up among rural people; as a result villagers became conscious of their rights and improved their standard of living.

Shortcomings and failure

1. Panchayat raj representative had a feeling that there is a dilution of their authorities and responsibilities in actual functioning.
2. There were cases which utilized directly the services of block level technical staff without involving the non-official.
3. Frequent transfer of staff gave poor image to the bodies.
4. Failure to mobilize resources to the extent necessary.

Important question

- Democracy word is derived from – Greek roots
- Demos meaning – 'The people'
- Kratos meaning – 'Authority'
- In democracy all authority originates from the – Peoples
- Devolution of central authority among local units close to the areas – Decentralisation
- Rajasthan became the pioneer to bring under democratic Decentralisation in – Oct. 2, 1959
- Define democratic Decentralisation
- Panchayat Raj System recommended by – B.R. Mehta Committee
- Three tier structure consists of – Zilla Parishad at district, Panchayat Samiti at Block and Gram Panchayat at Village level
- Three tier system of Panchayat raj was first adopted by and in – Nagaur, Rajasthan, October 2, 1959
- Write function of Gram Panchayat
- Draw extension organisation in Panchayat Raj Set-Up
- Write salient features of Panchayat Raj
- Explain the three tier of Panchayat Raj System

■■■■

12

Extension Administration

12.1 MEANING, CONCEPT

Concept of Agricultural Administration

Administration is usually connected with Organising, selecting executives or personnel, establishing plans, measuring results, coordinating, controlling and delegating activities (Ogunbameru, 2001) defined administration as giving direction and making decisions, formulating policies and plans, implementing the formulated policies and plans and monitoring them; maintaining inter personal relationships with others such as subordinates, those at the same level or peers, in other sectors, departments, organisation and superiors or super-ordinates using management techniques for planning, executing and evaluation of operatives; and providing administrative services such as clerical accounting, financial (FAO, 1979). Administration involves the creation and maintenance of an environment in which individual employees work together in groups towards the accomplishment of organisational goals.

Hardaker (1971), defined administration as forming balanced judgment, which for making rational decisions. A good administrator therefore implies acting with purpose, imagination, foresight and common sense.

Administration can also be seen as the act and science of getting result through employees. It is the marshalling or effective utilization of men, money, materials and methods (the four "M" of administrations). Agricultural administration is essentially the guidance, leadership and the control among all the clients of agriculture related activities (Ekpere, 1976). Efficient agricultural administration requires professionally trained and well-informed people. The chief role of the programme manager, managing director or director is to get results by effectively directing the work of other members of staff. To achieve this lofty role, the manager or administrator must adhere to the philosophy that hinges on:

1. Respect for the worth and dignity of the individual employees;
2. Faith in the employee's "ability to make rational decisions;
3. Maintain moral and spiritual values in inter-personal relationships;
4. Shared and collective responsibility for common goal of the organisation;
5. Respect for the dignity of labour; and
6. Promotion of the emotional, physical and psychological health of all employees, regardless of their tribe, religion, sex and political leaning.

Characteristics of Agricultural Administration

1. Agricultural administration demands an efficient organisation of the available manpower and physical resources with a view to ensure the greatest amount of goods for the largest number of people, in the least possible time and with minimum amount of energy.
2. A good agricultural administration always endeavours to be readily available and ever willing to solve the problems of the people as soon as they are brought to the notice of its personnel. It is also one of the tasks of agricultural administration to bring awareness among the people about the problems not readily perceived by them and to sensitize them to foresee their problems and be effective in resolving them.
3. Agricultural administration operates through understanding rather than dogmatic pressure, through building confidence

by offering assistance in helping the workers at the successive lower levels.

4. Agricultural administration allows enough freedom to its officials to discharge their responsibilities both efficiently and effectively. For this purpose, powers are delegated to the maximum extent and centralization of the administration is kept to the barest maximum. A good agricultural administrator is therefore ever willing to render assistance to anyone who needs it and is always there to set a good standard of service and cooperation, *i.e.*, leadership by example.
5. Agricultural administration helps to create and maintain an efficient organisation manned by well trained and qualified personnel with the proper bent of mind and aptitude so as to convey the latest technology and scientific innovations to the people speedily in such a manner that the people can put them into practice.
6. Decentralisation drives up the functions of decision and authorization to the top management it leaves action, when decided upon, to be carried out by the subordinate authority.
7. Coordination involves some supervision over the work of units, components, departments or divisions, but effective administration ensures that supervision does not lead to friction nor does it retard the growth of the lower level local authorities.

Objectives

At the end of this unit you should be able to,

- Explain the concept of Agricultural Administration.
- Enumerate the characteristics of Agricultural Administration

12.2 PRINCIPLES AND FUNCTIONS

Functions of an Extension Administrator (Extension Director): The Director of Extension is the organisation's Chief Executive. The function of the Director is to plan, organize, staff, direct and controls the activities of the organisation. The position includes responsibility for supervision, personnel management, salary administration, facilities management, fiscal management, programme development and co-ordination; programme execution and programme evaluation. The Director is ultimately responsible for managing the organisation and for giving it leadership and direction. The above functions must be delegated to some extent to others in the organisation. There is, also, a great need for effective communication within the organisation.

The Extension Director should have the following minimum qualifications

A. A university degree in a relevant field and, usually, a postgraduate degree, and

B. Several years of professional experience, preferably at different levels in the extension organisation

Other qualities expected of the Extension Director are

1. The ability to make decisions and follow through on those decisions
2. A positive attitude in dealing with superiors, subordinates, and the general public
3. A highly developed sense of co-operation
4. The ability to inspire in others a sense of responsibility and duty to the people being served
5. A basic understanding of the educational philosophy and methods of an effective extension organisation
6. The technical skills of management, and
7. An understanding of agriculture and the ability to relate to rural people

Other Responsibilities and Functions of an Extension Administration

1. *Delineation of the objectives of extension service* – since the extension service exists to fulfill set objectives, there is a central source from which activities are initiated

2. *Drawing broad plans of action* – he is responsible for drawing broad plans for restructuring activities in the extension service. The administrator must be fully conversant with the functions in other areas of agriculture.
3. *Assembling of various categories of staff* – provision of support staff at all levels is a major responsibility of top administrative staff. The top administrator should assemble an adequate working team of experts whose activities are complementary to each other.
4. *Delegation of function* – it is the usual practice for the chief executive to delegate functions. He should be able to assign specific functions to individual members of staff. In all cases, responsibilities are clearly written down and not merely stated verbally. Periodic review of job description is made by the administrator to reflect changing situations.
5. *Directing (supervising) delegated functions* – this responsibility assumes that the administrator has a clear understanding of his authority and how it is exercised. Directing and overseeing requires that an officer to whom function is assigned should report back to the administrator. The report could be oral or written.

Functions: POSDCORB

The functions of an administrator in the administrative process are:

a. Planning
b. Organising
c. Staffing
d. Directing
e. Coordinating
f. Reporting
g. Budgeting

13

Transfer of Technology

13.1 CONCEPT AND MODELS

Models of Training

Simple Model of Training Process

More effective behaviour of people –on-the-job –in the organisation is the primary objective of training process as well a whole. In the in the simple training process, improvement is dependent variable and participants and organisations independent variables.

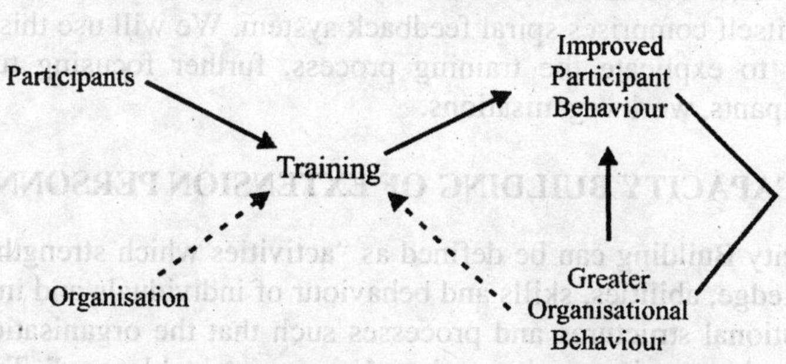

Elaborated Model of Training Process: Training is actually a more complex process. In the first place, the training system itself needs to be included. It may be temporary system, such as an occasional programme, or a permanent institution, such as training department. In either case, the trainers in the system also learn through the various opportunities available for checking their effecti-veness, *i.e.*

through feedback. Thus the independent and intervening variables become dependent variables.

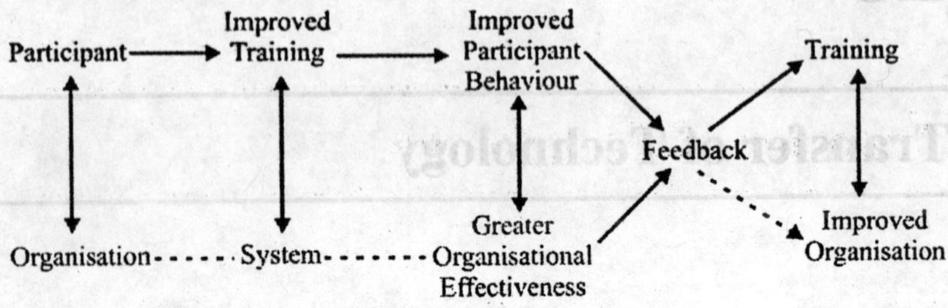

Elaborated Model of Training Process

Effectiveness

Spiral Model of Training Process

The phases, through which participants pass as they learn, then return (with enhanced capabilities, we hope) to their jobs. At various stages in the process the other two partners contribute "inputs" to assist the participants. These inputs are shown as arrows; arrows originating inside the spiral depict inputs of the work organisation; arrow originating outside the spiral, inputs of the training institutions. The arrows are merely visual conveniences. The large spiral itself comprises spiral feedback system. We will use this spiral model to explicate the training process, further focusing turn on participants, work organisations.

13.2 CAPACITY BUILDING OF EXTENSION PERSONNEL

Capacity Building can be defined as "activities which strengthen the knowledge, abilities, skills and behaviour of individuals and improve institutional structures and processes such that the organisation can efficiently meet its mission and goals in a sustainable way". Training is one of the essential components of capacity building.

Training for Development

Training has become an inseparable part of HRD. It has become one of the components, which enables any institution to churn out its employees as the most productive and most suitable ones.

Concept of Training: Lynton and Pareek (1967) stated that training consists of largely well organised opportunities for participants to acquire necessary understanding and skill. Training aims at lasting improvement on the job. The kind of education we all call as training is not for knowing more but behaving differently.

By Milton Mall (1980), Training is the process of aiding employees to gain effectiveness in their present or future work through the development of appropriate habits of thought, action, skills, knowledge and attitude.

Strayton (1986) defined training as acquisition and development of those knowledge, skills, techniques, attitudes and experiences, which enable an individual to make his most effective contribution to the combined efforts of the team of which he is a member.

Need for training

The process of training has caught up mainly in industries. This can be attributed to the sudden and competitive change that is occurring in the world. However, the needs for training can be fixed down to the following:

1. Rapid changes in technologies and jobs people do
2. Immediate and long term skill shortage
3. Changes in the expectation and composition of work force
4. Competition and market pressure for improvement in quality of products and services.

■■■■

14

Training

14.1 TYPES OF TRAINING

1. Pre-service Training

It is a process through which the individuals are made ready to enter a certain kind of professional job, as in agriculture, medicine or engineering. It is a professional training prior to any appointment, oriented to make an individual prepared to enter into a new profession. Swanson (1984) defines it as a programme of training activities that prepares an individual for a career in extension, and usually leads to some type of diploma, certificate, degree, or other qualification in one or more of the following agriculture, fisheries, forestry, animal and/or veterinary science or home science.

The state departments of Agriculture now prefer University graduates for entry into their extension services and similarly the Veterinary department prefers to take only Veterinary graduates released from the Universities.

2. In-Service Training

It is meant for in service candidates who are on the job. In-service training is a process of staff development for the purpose of improving the performance of an incumbent holding a position with assigned job responsibilities. It promotes the professional growth of individuals. Inservice training is a problem centred, learner oriented and time-bound series of activities, which provide the opportunity to develop a sense of purpose. Broaden perception of the participants and increase their capacity to gain knowledge and mastery of techniques.

According to Arnon (1987), even for the University graduate, learning cannot cease on completion of formal studies. He said that the in-service training is given with the following objectives

1. To keep up with research by regular meetings between researchers and extension workers, joint colloquia etc.
2. To impart basic knowledge not only in the fields directly related to agriculture, but also in sociology, economics, psychology etc.
3. To improve extension methods, by constant evaluation of methods, the joint study of research findings and extension methods, exchange of experiences.

In-Service training is of different types, some of them are as follows:

i. Orientation Training

This training is given usually to newly appointed extension personnel. It provides an introduction to public employment and provides answers to questions which a newly recruited person is likely to ask. This term is also used for training in-service extension personnel in a new responsibility likes a new operational programme so that personnel are appropriately oriented towards meeting the requirements of new situation.

ii. Induction/portal/vestibule Training

Induction training is given to new extension personnel immediately after they have been employed and before they are assigned to work in particular area usually as an Assistant Agriculture Officer or Agriculture Officer, or Extension Officer.

iii. Maintenance or refresher training

This training is originally started for trainers of the training institutes and Universities for refreshing their knowledge and skills for imparting them to trainees. The term indicates any new training for updating professional competence of extension personnel notably in the subject matter area of specialization. This training is usually imparted in the later career of extension personnel. This training is having considerable importance to extension personnel as it relates to updating the technical knowledge and competence of extension personnel. This deals with new information and new methods and review of older materials. This type of training is given to the employees to keep them at their peak performance level and also prevent them from getting into a rut.

iv. Retraining

It refers to the efforts designed to prepare an individual for a new assignment or a broadened aspect of the old specialty.

v. Career or development training / Training for professional qualification

This type of training is designed to upgrade the knowledge, skills and ability of employees to help them assume greater responsibility in higher positions. This training may lead to the acquisition of higher degree (undergraduate or postgraduate) or diploma by the employees, to motivate them to move up higher levels of administrative hierarchy (promotions). The Directorate of Extension is operating such a scheme on an yearly basis under which, in addition to salary and allowances which personnel get from their own employing organisations, it pays fixed monthly stipends to extension personnel to cover their cost of boarding, lodging and tuition fees. Only meritorious extension personnel and that too below the age of 45 years are eligible for such courses.

14.2 PLANNING TRAINING PROGRAMMES

Principals involved in Planning and Conduct of Training Programme

Training in an organisation is essentially a learning process in which learning opportunities the managerial purposefully structures, personnel and training staff, working in collaboration or by external agents acting on their behalf.

The aim of the process is to develop in the organisations employees the knowledge, skill and attitudes that have been defined as necessary for the effective performance of their work and hence for the achievement of organisational aim and objectives by the most effective means.

Components and sub-components of Training

Sl.No.	Component	Sub-Component
1.	Knowledge (K)	Knowledge of Extension Methods
		Subject Matter Knowledge
		Knowledge of Clients' Problems and their Solutions
		Knowledge of Farms, farm Families, and Farming Systems, and
		Knowledge of Farm Economics, Inputs Supply and Marketing
2.	Skills (S)	Communication Skills
		Technological Skills
		Demonstration Skills
		Organisational Skills
		Training Skills, and
		Diagnostic Skills
3.	Attitudes (A)	Attitude for Serving Clients
		Attitude for Acquiring Latest Knowledge and Technical Know-How
		Attitude for Problem-solving
		Attitude for Field Work, and
		Attitude for Self-Development
4.	Attributes (Os)	Personality Traits
		Conduct
		Empathy
		Experience, and
		Self-Development

Training Objectives

Any programme whether it be training or extension work should have objectives which are to be achieved. There are four types of objectives which are as follows:

Level	Type of objectives	What it should tell us
I	Policy/organisation	General direction of efforts, order of priorities
II	Target	How much is aimed
III	Operational objectives	Who has to do, what in order to attain the targets (in terms of clientele)
IV	Training objective	Who are the learners and what learning products are required

Principles involved in Training

Training should be based on the Principles of

1. Achievement of organisational objectives
2. Training need assessed
3. Training objectives – Expected learning behaviour
4. Flexibility
5. Level of the learners
6. Past experiences – existing knowledge
7. Co-operative/Collective process-active participation of learners
8. Teaching – learning process
9. Continuous process
10. Informed options – alternatives
11. Learning varies from individual
12. Individual creation of learning environment

Building up of a Training Programme

Any programme whether it be extension activity/Training activity, it has the following activities.

Trainer Roles

The various trainer roles can be seen in three distinct phases: pre-training, training and post-training.

a. Pre-training

1. Training Designer

The role of identifying and translating learning needs into objectives, content and designing the programme.

- Collecting and identifying learning needs
- Listing objectives
- Working out related contents/methods/materials/exercises
- Sequencing the contents/activities
- Identifying resource persons
- Preparing and selecting learning materials

2. Administrator/Organizer

The role of ensuring and meticulously planning in advance the facilities, learning materials, required equipment, participants and other related components of the training event and the coordination of the programme logistics.

- choosing venue and time
- selecting and scheduling facilities
- regularly communicating with the trainees regarding the programme plans
- identifying and arranging the needed support system at the training venue
- scheduling the time of co-trainees and resource persons
- distributing training materials
- arranging resources

b. During Training

1. Facilitator

The role of guiding the learning process so that individuals learn from each other and the group functions effectively.

- Eliciting opinions
- Enhancing participation
- Focusing trainees' attention on their potentialities
- Summarizing and synthesizing information
- Organising groups such that issues and needs are addressed
- Intervening in the process

2. Instructor

The role of presenting information and concepts, clarify objectives, creating and sustaining a structured learning environment and helping generate new learning.

- providing information and concepts
- directing structured learning – role-plays, simulations, games and discussions
- using learning aids – films, audio-tapes, video-tapes and other materials

3. Counselor

The role of supporting and guiding individual trainees during periods of stress and strain and helping trainees to assess their potentialities and personal competencies, so as to enable them to reflect, grow and change.

- Developing a rapport with trainees
- Showing genuine interest in directing their process of growth
- Communicating on a one-to-one basis
- Organising sessions to enhance self-confidence and self-esteem of some individuals

4. Recorder

The role of maintaining records of the process and content to enable monitoring, analysis and documentation.

- Observing keenly both flow of content and process
- Maintaining detailed notes on a daily basis

5. Evaluator

The role of assessing the impact of training programme on the trainees.

- Planning evaluation mechanisms
- Using written as well as verbal reports to assess an event
- Utilizing the evaluation design to assess individual changes in behaviour, attitudes and knowledge
- Forming steering committees to assist in day-to-day evaluations
- Conducting mid-term reviews
- Sharing reflections and analysis with co-trainer
- Providing relevant feedback

6. Organisers/Administrator/Manager

The role of managing all the related tasks during the programme

- Managing time and space for each session
- Solving problems related to accommodation, food, etc.
- Organising reservations, departures/arrivals, reimbursements, etc.
- Managing the learning situation (session timing, breaks, off-time, etc.)

c. Post-training

1. Report-writer

The role of preparing a report of the training programme.

- Organising the relevant information for the report-writing
- Disseminating the reports to all participants, and others interested

2. Follow-up Coordinator

The role of continuing contacts with individuals and their organisations to assess impact of training on the organisations and individuals, and providing the necessary follow-up support whenever needed.

- Communicating at regular intervals
- Inviting feedback from both organisations and individuals
- Collating learning needs for the next event, if so designed
- Providing support in the field

Important question related to topic

- Explain types of training.
- Write short notes on orientation, Induction and development training.
- Explain the component and sub-component of training.
- Describe training objectives.
- What are principles involved in training?
- As a trainer how to build up a training programme?
- What we do for pre and post-training?

■■■■

15

Diffusion and Adoption of Innovation

15.1 CONCEPT AND MEANING

An Innovation is an idea, practice or object that is perceived as new by an individual or other unit of adoption.

Diffusion is the process by which an innovation is communicated through certain channels over time among the members of a social system.

Adoption is a decision to make full use of an innovation as the best course of action available

15.2 PROCESS AND STAGES OF ADOPTION

According to Rogers, "adoption process is the mental process through which an individual passes from hearing about an innovation to final adoption", whereas diffusion process is the spread of a new idea from its source of innovation or creation to its ultimate users or adopters.

The adoption behaviour of an individual farmer has been conceptualized by researchers as a process composed of a number of successive stages. The five stage model of adoption process which is more commonly accepted than other models, is described below.

1. **Awareness:** The individual learns of the existence of the new idea, but lacks information about it. At this stage an individual is aware of the idea, but lacks detail information about it. For instance, the person may know only the name

and may not know what the idea is, what it will do or how it will work.

2. **Interest:** At interest stage a person wants more information about the idea or product. He wants to know what it is how it works and what its potentialities are. He may say to himself that this might help him increase his income, or help him control insects or diseases or improve farming or home life in some other way.

3. **Evaluation:** At this stage individual makes a mental trial of a new idea or practice. He applies the information obtained in the previous stages to his own situation. He asks himself: Can I do it? And if I do it, will it better than what I am doing now will it increase my income; or otherwise bring me satisfaction.

Stages in the Adoption Process

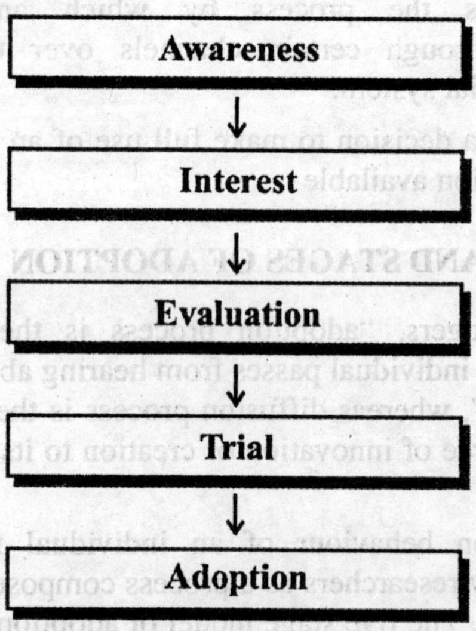

4. **Trial:** The individual actually applies the new idea on a small scale in order to determine its utility in own situation. If, in the judgment of individual, the innovation has some plus points *i.e.,* applicable to own situation, and if applied, shall in some way or other be of advantage, the person

takes a decision try it. These are generally small scale trials to test the effectiveness of the innovation in one's own situation. Apparently individuals need to test a new idea even though they have thought it for a long time and gathered information concerning it.

5. **Adoption:** The individual uses the new idea continuously on a full scale. Trial may be considered as the practical evaluation of an innovation. It provides evidence of the advantages of the innovation. Being satisfied with the trial and considering pros and cons of the situation, the individual takes a final decision and applies the innovation in a scale appropriate to own situation on a continued basis.

15.3 ADOPTER CATEGORIES

Adopter Categories

Definition of '**Adopter Categories**' classification of individuals based on their willingness to try out a new innovation or new product.

There are five categories of farmers. According to Rogers (1971), the farmers based on their innovativeness can be classified as

1. Innovators (2.5%, Venturesome)
2. Early adopters (13.5%, Respectable)
3. Early majority (34%, Deliberate)
4. Late majority (34%, Skeptical)
5. Laggards (16%, Traditional)

Characteristics of farmers

All individuals in a social system do not adopt an innovation at the same time. Rather, they adopt in an ordered time sequence, and they may be classified into adopter categories on the basis of when they first begin using a new idea. In technology transfer programme, it is of great practical utility for the extension workers to identify the individuals who are likely to adopt innovations early and who may lag behind.

The adoption of an innovation over time follows a normal, bell-shaped curve when plotted over time on frequency basis. If the cumulative number of adopters is plotted, it results in an 'S'shaped curve. The S-shaped curve rises slowly at first when there are few adopters in a time period, accelerate to a maximum when about half of the individuals in the system have adopted and then increases at a gradually slower rate as the few remaining individuals finally adopt . The S-shaped curve is like that of a 'learning curve' as propounded by the psychologists. Each adoption in the social system is in a sense equivalent to a learning trial by an individual.

Both of these curves are for the same data, the adoption of an innovation over time by the members of a social system. But the bell-shaped curve shows these data in terms of the number of individuals adopting each year, whereas the S-shaped curve shows these data on cumulative basis.

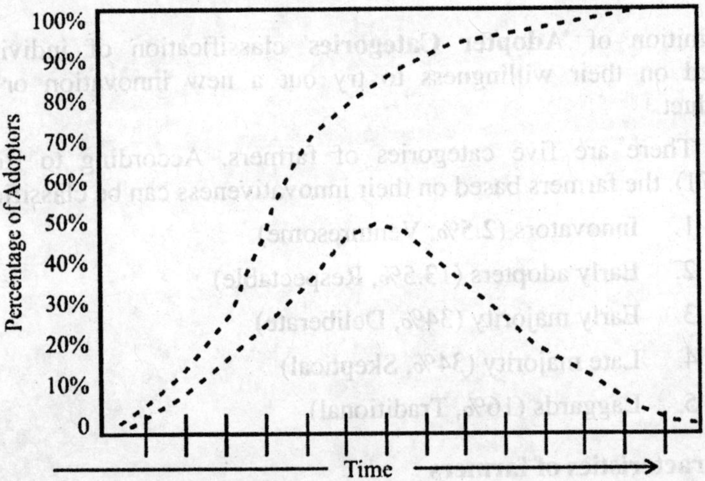

The bell shaped frequency curve and the S-shaped cumulative curve for adopter categories

The distribution of adopters over time closely approaches normaly, and may be explained by the statistical concept of normal curve. The distribution of the adopters may be partitioned into five adopter categories by using the mean (x) and standard deviation. The area lying to the left of the mean time of adoption minus two standard deviations includes 2.5 per cent of the individuals who are the first to adopt an innovation and are known as innovators. The next 13.5 per cent between the mean minus one standard deviation and the mean minus two standard deviations to adopt the new idea are called as early adopters. The next 34 per cent of the adopters between the mean date of adoption and minus one standard deviation are known as early majority. Between the mean and one standard deviation to the right of the mean are located the next 34 per cent to adopt the new idea, the late majority. The last 16 per cent to the right of mean plus one standard deviation are the last to adopt the innovation the laggards. The five-adopter categories are conceptualized as ideal types and are presented in below diagram.

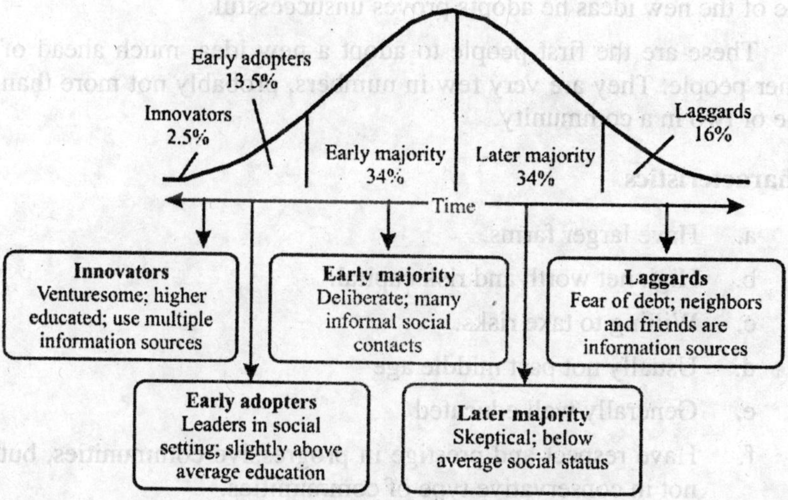

The innovativeness dimension, as measured by the time at which an individual adopts an innovation, is continuous. However, this variable may be partitioned into five adopter categories by lying of standard deviations from the average time of adoption.

The detailed information on the characteristics of adopter categories is presented in the succeeding pages.

Innovators: Venturesome

Observers have noted that venturesomeness is almost an obsession with innovators. They are eager to try new ideas. This interest leads them out of a local circle of peers and into more cosmopolite social relationships. Communication patterns and friendships among a clique of innovators are common; even though the geographical distance between the innovators may be great. Being an innovator has several prerequisites. These include control of substantial financial resources to absorb the understand and apply complex technical knowledge.

The salient value of the innovator is venturesomeness. He desires the hazardous, the rash, the daring, and the risky. The innovator also must be willing to accept an occasional setback when one of the new ideas he adopts proves unsuccessful.

These are the first people to adopt a new idea, much ahead of other people. They are very few in numbers, probably not more than one or two in a community.

Characteristics

a. Have larger farms.

b. High net worth and risk capital.

c. Willing to take risks.

d. Usually not past middle age

e. Generally well educated

f. Have respect and prestige in progressive communities, but not in conservative type of communities.

g. Mentally alert and actively seeking new ideas.

h. Their sphere of influence and activity often goes beyond the community boundaries.

i. They have much formal and informal contact outside the immediate locality.

j. They often by-pass the local extension worker in getting information from the originating sources, and may learn about new things even before he does. They sometimes manage to get samples of seeds or chemicals even before they are released for public use.

k. They subscribe to many farm magazines and specialised publications.

l. Other farmers may watch the innovators and know what they are doing, but the innovators are not generally named by other farmers as "neighbours and friends" to whom they go for information.

Early Adopter: Respectable

Early adopters are more integrated part of the local social system than innovators. Whereas innovators are cosmopolites, early adopters are localities. This adopter's category, more than any other, has the greatest degree of opinion leadership in most social systems. Potential adopters look to early adopters for advice and information about the innovation. The early adopter is considered by many as "the man to check with" before using a new idea. This adopter category is generally sought by change agents to be a local missionary for speeding the diffusion process. Since early adopters are not too far ahead of the average individual in innovativeness, they serve as a role model for many other members of a social system. Members of a social system respect the early adopter. The early adopter is respected by his peers. He is the embodiment of successful and discrete use of new ideas. And the early adopter knows that he must continue to earn this esteem of his colleagues if his position in the social structure is to be maintained.

Characteristics

a. Younger than those who have a slower adoption rate, but not necessarily younger than the innovators

b. They are not the persons who test the untried ideas, but they are quickest to use tried ideas in their own situations.

c. Have large farms.

d. Higher education than those who adopt more slowly.

e. High income.

f. They participate more in the format activities of the community.

g. They also participate more in government programmes.

h. This group usually furnishes a disproportionate amount of the formal leadership (elected positions) in the community.

i. They read papers and farm journals and receive more bulletins than people who adopt later.

j. They may be regarded as community adoption leaders.

Early Majority: Deliberate (Local Adoption Leaders)

The early majority adopt new ideas just before the average member of a social system. The early majority interacts frequently with their peers, but leadership position is rarely held by them. The early majority's unique position; between the very early and relatively late to adopt, make them an important link in the diffusion process. The early majority may deliberate for some time before completely adopting a new idea. Their innovation-decision is relatively longer than that of the innovator and the early adopter. "Be not the last to lay neither the old aside, nor the first by which the new is tried", might be the motto of the early majority. They follow with deliberate willingness in adopting innovations, but seldom lead.

Characteristics

a. Slightly above average in age, education and farming experience.

b. They take a few more farm journals and bulletins than the average.

c. They have medium high social and economic status.

d. Less active in formal groups than early adopters, but more active than those adopting later.

e. In many cases, they are not formal leaders in the association

f. They also attend extension meetings and farm demonstrations.

g. They are most likely to be informal resources than early adopters and innovators, and so cannot afford to make hasty or poor decisions.

h. They associate mainly with people of their own community.

i. They value highly the opinions their neighbours and friends hold about them, for this is their main source of status and prestige.

j. They are mostly mentioned as "neighbours and friends" from whom the majority of farmers seek information.

Late Majority: Skeptical

The late majority adopt new ideas just after the average member of a social system. Adoption may be both an economic necessity and the answer to increasing social pressures. Innovations are approached with a skeptical and cautions air, and the late majority do not adopt until most other in their social system have done so. The weight of system norms must definitely favour the innovation before the late majority is convinced. They can be persuaded of the utility of new ideas, but the pressure of peers is necessary to motivate adoption.

Characteristics

a. Those in this group have less education and are older than the early majority.

b. They form the major part, 9 formal organisational membership, although they participate less in such formal groups.

c. They take fewer leadership roles than the earlier adopters.

d. They take and read fewer papers, magazines and bulletins, than the early majority.

e. They do not participate in as many activities outside the community as do people that adopt earlier.

Laggards: Traditional

Laggards are the last to adopt an innovation. They possess almost no opinion leadership. They are the most localities in their outlook of all adopter categories, many are near isolates. The point of reference for the laggard is the past. Decisions are usually made in terms of what has been done in previous generations. This individual interacts primarily with others who have traditional values. When laggards finally adopt an innovation, it may already have been superseded by another more recent idea which the innovators are already using. Laggards tend to be frankly suspicious of innovations, innovators, and change agents. Their tradition direction slows the innovation decision process to a crawl. Adoption lags far behind knowledge of the idea. Alienation from a too-fast-moving world is apparent in much of the laggard's outlook. While most individuals in a social system are looking to the road of change ahead, the laggards has his attention fixed on the rear-view mirror.

Important questions

- Define An Innovation, Diffusion and Adoption
- Explain process and stage of adoption
- Describe the Adopter categories as per Roger's
- Innovator is also known as – Venture some, 2.5%
- Early Adopters – Respectable, 13.5%
- Early Majority – Deliberate, 34%
- Late majority – 34%, Skeptical
- Laggard – 16%, Traditional
- Describe the Attributes of Innovation.

16

Questions for the Practice

- Extension aims for increase in…………..Skill and Attitude of the farmers
- Communication is a process by which two or more people **exchange** ideas, facts, feelings and impression.
- The telephone number of Kisan Call Centre is **1551**
- The book entitles "Extension Education" is written by **A.A. Reddy**
- It is estimated that according to innovativeness only 2.5 percent farmers are **Innovators.**
- Full use of innovation is called **Adoption**
- Audio cassettes are **Mass contact** teaching methods
- The first step of programme planning is collection and **Analysis** of facts
- First model of communication was given by **Aristotle, 1960**
- **Feedback** is the response of audience back to the communicator
- On the basis of **innovativeness** farmers can be grouped in five categories.
- Communication is derived from the **Latin** word **communis** which means **common**
- Video conferencing is done through **Mobile and Computer**

- Video tapes are most **important Mass media** in transfer of technology
- AIDCAS was given by **Wilson**
- Extension teaching methods **AIDCAS**
- **Extension teaching** method is a face to face individual contact methods
- Predetermined course of action **Plan**
- Objectives are **expression** of the ends towards which our efforts are directed
- Direction of movement **objectives**
- The communicators **message** should make the intended impression on the receiver
- Calendar of work is a plan of work arranged **Activity**
- Programme should have balance with **Emphasis**
- Adoption is a **continuous** and **full** use of an idea
- Diffusion is a part of **communication** process
- **Channel** are the devices used to create situations in which communication take place between source and receiver
- **Panel** is a short series of lecture, usually by two to three speakers
- A,B,C of poster are **attitude, brief and clear**
- Channels are the physical bridges between the **communicator** and **receiver**
- A **message** is the information a communicator wishes his audience to receiver, understand accept and act upon
- Fidelity is the faithful performance of communication process by its all elements.
- **Planning** is a comparison of the situation, before and after a development programme, has operated within it for a pre-determined period.
- Method demonstration is useful in developing **confidence**
- Extension programme must be based on **Need** of the peoples

- Seminar is a **Group** discussion technique
- Farm clinic is a **individual** contact method
- **Rejection** is a decision not to adopt an innovation
- Diffusion is the process by which an **innovation** is spread to the members of social system
- Krishi Vigyan Kendra are sponsored by **ICAR**
- T & V System is also called……….. Scheme
- Firka development scheme was started in the year …………
- ……………….. is the third tier of the Panchayat raj, operating at the district level
- Community development programme was started in……………
- Democratic decentralisation was recommended by a committee
- The two employment programme, NREP and RELGP have been merged
- A three-tier system of Panchayat raj was introduced in the country with the operation of which constitutional amendment
- ATMA is a…………….. society
- ATIC are set up by
- In T & V system, the number of farmers in each group is ……
- National demonstration was started in the year
- Extension basically is a two-way process between whom?
- Village Panchayat is a non-official body
- Extension programme should started with local group, local situation and local problem
- The word extension originated from the language …Latin
- The fundamental objectives of extension is the …….of people
- Sriniketan project was started in the year ....
- NES programme was started in the year ....
- ……….. Committee recommended the imitation Panchayat raj system in our country

- Community development programme was started in the year…….
- ………Amendments associated with Panchayati raj system.
- NATP was started in the year..
- ATMA stands for……..
- T&V extension system was first introduced on a pilot basis in the Rajasthan Canal and ………….. area in MP.
- Extension education and its aim is to bring the…….in human behaviour.
- The general body of village Panchayat is……….
- Mohan Singh Mehta committee was appointed in 1974 by ICAR to formulate……………..scheme.
- ATIC is intended to provide all basis needs of the farmers through ……….. services.
- Community development programme was started in the year …………
- The full form of IVLP is………………
- The basic concept of extension is that it is…………
- Democratic decentralisation was recommended by a committee headed by…….
- Education is a production of desirable change in human……..
- The concept of Panchayati raj system was introduced in India on the recommendation of…………….
- National extension service was started in the year………
- The term 'Extension 'has been derived from……. language.
- The fundamental objective of extension is…………..
- NATP was started in the year…………

MCQ's

1. ATIC are set up by

A. Ministry of Agriculture, Govt. of India

B. State Department of Agriculture

C. INCAR

B. Ministry of Rural Development,GOI

2. In T&V system, the number of farmers in each group is

A. 6 B. 8

C. 10 D. 12

3. National Demonstration was started in the year

A. 1975 B. 1974

C. 1963 D. 1964

4. Extension basically is a two-way process whom

A. Farmer and Researcher

B. Researcher and Planner

C. Teacher and planner

D. Researcher and Teacher

5. Basic principle of extension education

A. Help those who help themselves

B. Internal and spiritual education

C. Help poorest of poor

D. Adult education

6. The best example of democratic Decentralisation is:

A. Parliament B. Panchayati Raj

C. Legislative assembly D. None of the above

7. The author of a book 'Fxtension Education I' is

A. G.L. Ray B. B.B. Mohaty

C. A. Adivi Reddy D. Edgar Dale

8. The second step of extension education process is:
   A. Formulation of objectives
   B. Evaluation of programme
   C. Analysis the situation
   D. Developing plan of work
9. In which project, first time emphasis was laid on curtailing expenditure on social and religious functions.
   A. Marthandam B. Gurgaon
   C. Sevagram B. Sriniketan
10. Etawah pilot project was started in the year
   A. 1945 B. 1952
   C. 1948 D. 1950
11. Education object is:
   A. Change in knowledge B. Change in attitude
   C. Change in Skill D. Change in all above
12. The first KVK of Rajasthan was establishing in the district:
   A. Sikar B. Bharatpur
   C. Jalore D. Chittorgarh
13. The total of KVK in Rajasthan at present are:
   A. 40 B. 43
   C. 42. D. 44
14. To which programme name of S.K. Dey is associated:
   A. Democratic Decentralisation
   B. KVK
   C. Nilokheri
   D. Marthandam programme
15. The term 'Extension Education' was first used in the year
   A. 1866 B. 1873
   C. 1878 D. 1979

16. Lab to Lab land programme was started in the year

A. 1952 B. 1953

C. 1966 D. 1979

17. Extension Education is a

A. Behavioural process B. Natural Science

C. Physical Science D. Earth Science

18. Where did Sir Daniel Hamilton start scheme of rural reconstruction?

A. Sundarban B. Pohari

C. Wardha D. Nagpur

19. In which district of Rajasthan Raj System was started for the first time in India ?

A. Jaipur B. Bikaner

D. Nagpur D. Udaipur

20. National Extension Service was started in the year

A. 1952 B. 1953

C. 1954 D. 1954

21. Sevagram programme was initiated by

A. F. Nicholson B. Ravindra Nath

C. Mahatma Gandhi D. S.K. Dubey

22. Which programme was started by F.L. Brayne?

A. Sarvodaya programme

B. Sriniketan

C. Sevagram D. Gurgaon project

23. Etawah Pilot project was started by

A. Spencer Hatch B. Hatch

C. F.L. Brayne D. Albert Mayor

24. Changes in human behaviour include?

A. Change in Knowledge B. Change in Skill

C. Change in Attitude D. All of the above

25. Basic Principle of extension education is

A. Adult education

B. Help poorest of poor

C. Helping people to keep themselves.

D. Moral education for farmers

26. The best example of 'Democratic Decentralisation is:

A. Parliament B. Legislative Assembly

C. Panchayat Raj D. None of the above

27. Establishment of Krishi Vigyan Kendra was recommended by:

A. Dr. Ashok Mehta committee

B. Dr. Manmohan Singh Mehta committee

C. Dr. Balwant Rai Mehta Committee

D. Dr. Mohan Singh Mehta Committee

28. Agricultural Technology information Centre in the country are set up by:

A. Ministry of Rural Development, GoI

B. Ministry oh HRD, GoI

C. ICAR, New Delhi

D. Ministry of Agriculture, GoI

29. NATP was started in the financial year

A. 1977–98 B. 1995–96

C. 1991–92 D. 1990–91

30. 'Gurgaon project' was started by

A. Spencer Hatch B. F.L. Brayne

C. Albert Mayer D. R.N. Tagore

31. 'Marthandam programme' was implemented in;

A. Punjab B. Uttar Pradesh

C. Haryana D. Tamil Nadu

32. Nilokheri Rural Development programme was started by

A. Jawahar Lal Nehru B. F.L. Brayne

C. Mahtma Gandhi D. S.K. Dey

33. The first KVK of the country was established in the year

A. 1997 B. 1974

C. 1976 D. 1977

34. Which one of the following is project teaching aid

A. Slide B. Audio Cassettes

C. Chart D. Folder

35. Behvioural change of human being in Extension education includes change in

A. Knowledge B. Attitude

C. Skill D. All the above

36. How much percentage of farmers in a social system is 'late Majority Adopter?'

A. 15 B. 16

C. 13.5 D. 34

37. Which type of teaching aid video cassettes are?

A. Audio B. Visual

C. Audio-Visual D. Non-project

38. Which one of the following is not an element of diffusion process?

A. Innovation B. Social System

C. Communicator D. Over time

39. Which one of the following is not an element of communication process?

A. Message B. Source

C. Teaching aid D. Receiver

40. Puppet is a

A. Individual contact method B. Group contact method

C. Mass contact method D. Project teaching aid

41. Which one of the following is not a stage of adoption process?

A. Interest B. Skill

C. Awareness D. Trial

42. The early adopter farmers are also called

A. Respectful B. Deliberate

C. Venturesome D. Skeptical

43. The gap between "what is" and "what ought to be" is called

A. Attitude B. Need

C. Behaviour D. Aspiration

44. Communication process is a

A. One-way process B. Two-way process

C. Three- way process D. None of these

45. Which of the following is not considered as mass media?

A. Radio B. Newspaper

C. Television D. Personal letter

46. Which of the following is not the project visual aid?

A. Film B. Slide

C. Flash D. None of these

47. Communication process never completes without

A. Feedback B. Audience response

C. Communicator D. All the above

48. Which of the following is not a way of communication?

A. Circular Communication

B. Diagonal Communication

C. Interpersonal Communication

D. Horizontal Communication

49. Noise is a communication barrier related to

A. Communicator B. Channel

C. Message D. Receiver

50. Which of the following is not a element of communication process?

A. Communication sources B. Communication fidelity

C. Communication feedback D. None of these

51. An outline of activities so arranged as to enable execution of the entire programme is known as

A. A project B. A calendar of work

C. A plan of work D. None of these

52. Of the following which is not the individual contact method?

A. Farm and Home visit B. Office call

C. Result Demonstration D. Circular letter

53. While adoption of an exotic breed cow which step is generally skipped?

A. Awareness B. Interest

C. Trial D. Evaluation

54. When there is limited time for discussion, the large group is divided into smaller units is called

A. Buzz Session B. Brain storming

C. Seminar D. Conference

55. How many innovators are found in the community

A. 16% B. 2. 5%

C. 13.5% D. 34%

56. Encoder and Decoder are the elements of

A. Leagan's model B. Berol's Model

C. Litter's Model D. Nucomb Model

57. Answer the questions how, when, where and by whom the work is to be done?

A. Programme

B. Plan of work

C. Project

D. Objective

58. **Informal conversation put on for the credit of a audience by a small group of speakers, usually 2 to 8 in number, is called**

A. Seminar B. Panel

C. Forum D. Workshop

59. Formation of attitude is related with

A. Decision B. Persuasion

C. Confirmation D. Action

60. Of the following which is not the element of diffusion process?

A. Innovation B. Communication

C. Overtime D. Implementation

61. Of the following which is not the mass contact method?

A. Radio B. Circular letter

C. Tours D. Poster

62. Which one of the following aids involves two sensory organs simultaneously?

A. Model B. Poster

C. Television D. Radio

63. Symposium is a short series of lecture, usually by

A. 1 to 2 Speakers B. 6 to 8 Speakers

C. 2 to 5 Speakers D. None of the above

64. In the following which one of the individual contact method?

A. Method demonstration B. Field day

C. Group Discussion D. Office call

65. The Leagan's model of communication contains

A. Six elements B. Eight elements

C. Five elements D. Three elements

66. Communication involves three phases, *i.e.,* expression, interpretation and

A. Response B. Transmission

C. Adoption D. None of the above

67. **Plan of work include (When, Where and Whom)**

A. Four w's B. Three w's

C. Five w's D. None of the above

68. **In the following which is not the elements of diffusion process?**

A. Innovation B. Farmers

C. Evaluation D. Overtime

69. **Adopter's categories are**

A. Five B. Eight

C. Seven D. Ten

70. **The first step of adoption process is**

A. Adoption B. Interest

C. Awareness D. Evaluation

Definitions

1. Democratic decentralisation **2.** Formal education **3.** Education **4.** Non-formal education **6.** NES Extension Education **7.** Extension Services **8.** Extension Process **9.** Lecture **10.** Rate of adoption **11.** Over adoption **12.** Innovation **13.** Community **14.** Objective **15.** Attitude **16.** Informal Education **17.** Marthandam project **18.** IVLP **19.** TRYSEM 20. Programme **21.** Planning **22.** Project **23.** Teaching methods **24.** Diffusion **25.** Adoption **26.** Sriniketan Development programme **27.** ATMA **28.** MNREGA **29.** Feedback **30.** Evaluation **31.** Goal **32.** Innovations **33.** Treatment of message **34.** Encoding and Decoding

Short notes

1. Marthandam project **2.** Etawah pilot project **3.** Sriniketan project **4.** NES **5.** Exhibition **6.** Demerits of mass contact method **7.** Step of adoption **8.** Relationship between communication and diffusion **9.** Elements of Leagan's model communication **10.** Enlist barrier of communication **11.** Degree of evaluation **12.** Explain characteristics of Innovation **13.** Discuss social factors affecting adoption of innovation **14.** Elements of communication **15.** Formal and Informal evaluation **16.** Key elements of Non-Formal education **17.** Difference between Formal education and Extension Education **18.**

Objectives of Extension Education **19.** Model of communication **20.** Mass communication **21.** Objectives of KVK **22.** ETAWAH Pilot Project **23.** Difference between SGSY and IRDP **24.** Panchayati Raj System **25.** Concept of community development programme **26.** Objectives of NAIP **27.** Principles of KVK **28.** Adopter categories **29.** Factors influencing adoption process **30.** Method demonstration **31.** Barrier in communication **32.** Element of diffusion **32.** Difference between CDP and IRDP **33.** Differentiate the agricultural extension and Extension Education **34.** Objectives of extension education **35.** Organisational Structure of Panchayati Raj System **36.** Objectives and Component of Agriculture Technology Information Centre **37.** Three-tier system of Panchayat raj **38.** What is T&V system? Explain its key features and limitations **39.** Advantage of Democratic Decentralisation **40.** Differentiate between formal and Non-formal education **41.** Objective and Goal **42.** Write the role of Etawah pilot project in rural development **43.** Draw the organisational structure of Community Development Project. **44.** Draw organisational set up T&V system and write important characteristics of the system **45.** Write the key features of Gurgaon project **46.** Discuss IVLP in detail with special reference to its objectives and salient features. **47.** Explain at length the principles of Extension Education **48.** Discuss the objectives and operational methodology of SGS **49.** Describe the role of Panchayati Raj System in Rural Development in Detail. **50.** Define Community Development Programme and write it's steps and objectives **51.** What is Training and Visit System Explain its Objectives and Methodology **52.** What is the result demonstration? Explain the steps of conducting result demonstration **53.** Discuss the elements of diffusion process **54.** Discuss the factors affecting selection and use of extension teaching methods **55.** Discuss the principles and objectives of Extension Education in detail **56.** Describe in detail the Agricultural Technology Management Agency **57.** Discuss any two development programme of Pre-Independence era with reference to objectives and activities. **58.** What do you mean by adoption? Discuss in detail five stage model of adoption process. **59.** Explain in detail the factors influencing selection of extension teaching method **60.** What do you understand by evaluation? Explain in detail the degree (types) of evaluation. **61.** Describe the importance of Organising 'Kisan Mela'. Explain its planning and execution

62. What is principle? Enlist the principles of programme planning and prepare a programmed for improvement of crop production system under rainfed condition. **63.** Explain adopter's categories in a social system on the basis of innovativeness **64.** Discuss the factors affecting adoption of innovation. **65.** Describe the principles of programme planning **66.** What is the method Demonstration? Explain the steps of conducting method demonstration.

■■■■

17

Human Values and Ethics - An Introduction

Human values are the virtues that guide us to consider the human element when one interacts with other human beings. They are our strong positive feelings for the human essence. Values said to be 'ethical' are those that command respectful behaviour towards others, which is towards other human beings, without harming them.

The human ethical and moral values are universal values. They are also expressed formally in laws, constitutions towards asserting human rights. Human values inspire us to practice our moral values, especially in conflict situations, when we tend to put them aside and become confrontational.

A Value system's an enduring organisation of beliefs concerning preferable modes of conduct along a continuum of importance. Thus the importance of different values co-varies with the importance of others in the value system, for *e.g.*, one may value 'honesty over successes'.

17.1. DEFINITIONS OF HUMAN VALUES

- A value is defined as a principle that promotes well-being or prevents harm
- Values are guidelines for our success, our paradigm about what is acceptable

- Human values thus defined are universal: they are shared by all human beings, whatever may be their region, nationality, culture, or personal history. By nature, human values induce consideration for others

Some example of human values

- Brotherhood, friendship, empathy, compassion, love
- Openness, listening, welcoming, acceptance, recognition, apperception.
- Honesty, fairness, loyalty, sharing, solidarity
- Civility, respect, consideration

Human values – values are guiding principles of our life

Now they are essential for human behavioural or action in our daily life.

- There formal basis of interest, choice, need, desire and preference
- They have important role in sociology

Why values needed

1. Guide use in right path
2. Learn of important certainty, goodness and verify
3. Give direction to life being joy
4. Learn satisfaction towards life
5. Attent to peace of life of culture and heritage

They being change in behaviour towards positive thoughts.

1. Intrinsic value, *e.g.*, of Intrinsic value is happiness, joy etc.
2. Extrinsic values, of extrinsic values –Truth, honesty etc.

Characteristic of values

1. Live
2. Value is refereed to desired goal and emotional action
3. Value is standard and criteria

4. Value are ordered by important to one-another, relative important multiple value action.

Ethics: Branch of philosophy that reads to study ideal human behaviour and an ideal wave of being.

Ethical and unethical judge by social stander are varies person to person

Goal: A goal is defined as a target an dream is a action with a purpose goal will unable for fulfill a purpose live for dream.

Importance of Goal

1. Goal motivates you
2. It makes decision easy
3. To be health
4. It build confidence
5. Show your attitude
6. It gives direction and purpose
7. It gives real meaning of activity

Evolution of Human Values

The human values evolve because of the following factors:

1. The emotions developed or modified by one's own awareness, choice, and judgment in fulfilling the needs.
2. The teachings and practice of Preceptors (Gurus) or Saviors or Religions leaders.
3. Fostered or modified by social leaders, rulers of kingdom, and by law (government). Pact of norms of the society on the fulfillment of the individual's needs or desires

Respect – It is one of the most important human value for establishing relations of peace and yet it remains elusive. Its understanding varies according to age (child, teen, adult), to one's education and surrounding culture. Indeed, to respect someone, one must be able to appreciate some of his/her human qualities, even if one does not appreciate his/her opinions or past behaviour.

17.2. TYPES OF VALUES

The five core human values are

1. Right conduct
2. Peace
3. Truth
4. Love
5. Nonviolence
6. Values related to Right Conduct are:
   - (a) Self-help skills: Care of possessions, diet, hygiene, modesty, posture, self reliance, and tidy appearance.
   - (b) social skills: Good behaviour, good manners, good relationships, helpfulness, no wastage, and good environment, and
   - (c) ethical skills: Code of conduct, courage, dependability, duty, efficiency, ingenuity, initiative, perseverance, punctuality, resourcefulness, respect, for all, and responsibility

2\. Values related to Peace: Attentions, calmness, concentration, contentment, dignity, discipline, equality, equanimity, faithfulness, focus, gratitude, happiness, harmony, humility, inner silence, optimism, patience, reflection, satisfaction, self-acceptance, self-confidence, self-control, self-discipline, self-esteem, self-respect, sense control, tolerance, and understanding

3\. Values related to Truth: Accuracy, curiosity, discernment, fairness, fearlessness, honesty, integrity (unity of thought, word, and deed), intuition, injustice, optimism, purity, and quest for knowledge, reason, self-analysis, and sincerity, spirit of enquiry, synthesis, trust, truthfulness and determination.

4\. Values related to Love: Acceptance, affection, care, compassion, consideration, dedication, devotion, empathy, forbearance, forgiveness, friendship, generosity, gentleness, humanness, interdependence, kindness, patience, patriotism, reverence, sacrifice, selflessness, service, sharing, sympathy, thoughtfulness, tolerance and trust

5. Values related to Non-Violence are:

   a. Psychological: Benevolence, compassion, con-cern for others, consideration, forbearance, forgive-ness, manners, happiness, loyalty, morality, and universal love

   b. Social: Appreciation of other cultures and religions, brotherhood, care of environment, citizenship, equality, harmlessness, national awareness, perseverance, respect for property, and social justice.

Perseverance – It is defined as persistence, determination, resolution, tenacity, dedication, commitment, constancy, steadfastness, stamina, endurance and indefatigability. To persevere is described as to continue, carry on, stick at it (in formal), keep going, persist, plug away (informal), remain, stand firm, hold on hang on, Perseverance builds character.

Accuracy – It means freedom from mistake or error, conformity to truth or to a standard or model and exactness. Accuracy is defined as correctness, exactness, authenticity, truth, veracity, closeness to truth (truth value) and carefulness. The value of accuracy embraces a large area and has many implications. Accuracy includes telling the truth, not ex-aggerating, and taking care over one's work.

Discernment – It means discrimination, perception, penetration, and insight. Discernment means the power to see what is not obvious to the average mind. It stresses accuracy, especially in reading character or motives. Discrimination stresses the power to distinguish or select what is true or genuinely excellent. Perception implies quick and often sympathetic discernment, as of shades of feelings. Penetration implies a searching mind that goes beyond what is obvious or superficial. Insight suggests depth of discernment.

Role of family and society in inculcating values

- The family and society is important in developing the moral values of a child. There is a close contact between the parents and children. Which determine the personality of child. Family is the foundation on which values are built.

- Moral values like truthfulness, happiness, peace, justice are instilled in children's thoughts, feelings and actions, and they function as ideals and standards that govern their actions in their life. The value system practiced in the family becomes automatic to the young family members if they are taught moral values systematically.
- The family, shapes the child's attitude towards people and society, and helps in mental growth in the child and supports his ambitions and values. Blissful and cheerful atmosphere in the family will develop the love, affection, tolerance, and generosity. A child learns his behaviour by modeling what he sees around him.
- Family plays a major role in helpings a child socialize and has great influence and bearing on the progress of the child. Joint family system, the presence of elders in the family, plays the effective role in social and moral development of the children. It will also help young generation of the family to Imbibe human values and eradicate their negative mental tendencies when they are among elders.
- Children indentify themselves with their parents, other family elders and adopt them as their personal models for emulation and imitation. The behavioural problems are set correct only by the involvement of family in the child's life as they spend most of their time in adolescence with parents.
- Family is the first social organisation that provides the immediate proximity from which the kid can learn his behaviour.
- Social standards and customs defined by a family provide the emotional and physical basis for a child. Values developed by a family are the foundation for how children learn, grow and function in the world. These beliefs, transmits the way of life a child lives and changes into and individual in a society. These values and morals, guide the

individual every time in his actions. Children turn out to be good person because of the value taught and given by their family members. Ideas passed down from generation to generation make up a family values. Customs and traditions followed and taught by the family leads a disciplined and organised life.

- Family values help the child to stand strong on his views despite efforts by right and wrong and are less likely to become victims of deviant influences.

Role of educational institutions in inculcating values

In school, children are members of a small society that has a great influence on their moral development. Teachers serve as role model for students in school. They play a major role in developing ethical behaviour in children. General Steps are:

- *Accountability:* The children should be encouraged to be accountable for their own actions and should learn to respect and treat others kindly.
- *Role model:* The teachers are the first role model to the children outside their family. When the children see the model showing concern for others, motivating them for their good deeds and cooperating and helpful with their academic issue, the children learn them by observing and imitate it with fellow peers.
- *Helping:* The children are taught basic morals and values in school. They should be taught by emphasizing the idea through many activities, stories and tales, which will encourage them to engage in more helping behaviours.
- *Appreciation:* The teacher should appreciate the children for developing pro-social behaviour, especially for any specific action they have done to help others.

17.3. VALUE EDUCATION

Value means importance and education means knowledge, therefore value education means knowledge of importance. In other words when we get the knowledge of importance of anything, importance of human being, importance of us as human being, then we have value education.

Value education deals with what is universally valuable to us, what is conductive to our individual and collective happiness and prosperity in a sustainable way.

Value education is defined as the process that gives young people an initiation into value of the rules relating to other people, a grasp of certain underlying principles, together with the ability to apply these rules intelligently.

Significance of Value Education

- Education is important to help everyone in improving the value system that he/she puts to use.
- Understanding of one's own values, helps an individual to examine and control the various choices to be made in life. One has to frequently uphold the various types of values in life, such as cultural values, universal, personal values and social values.
- Value education is always essential to shape one's life and to give one an opportunity to perform on the global stage. The need for value education among the parents, children, teacher etc, is constantly increasing as we continue to witness increasing violent activities, behavioural disorders and lack of unity in the society etc.
- Value education enables us to understand our needs and visualize our goals correctly and it also indicates the direction for their fulfillment.
- It also helps to remove our confusions and enables us to rightly utilize the technological innovations.

The Role of Value Education

Value education plays a very important role in creating a better society, more ethical organisations

The rule of value education is as follows:

1. Value education helps to build human beings who possess strength, integrity and fortitude.

2. Value education inculcates cooperation, peace as well as tolerance.
3. Efficiency can step up if a person possesses the right values. This may include punctuality, keeping promise, professionalism, lack of bias or prejudice etc.
4. Creating cordial relationship among people by encouraging the value respect, love and affection.
5. Promoting personality development and social cohesion.
6. Regenerating values of national pride and integration towards nation building.
7. Character building in the young people who are future leaders.
8. Promoting harmony between nations and creating a peaceful world order.
9. Identifying the core universal values like truth (Satya), Righteous Conduct (Dharma), peace (Shanti), Love (Prema) and Non-Violence (Ahimsa)
10. Help to create a foundation of the quality of life and strike a balance between external and internal values.

Thus value education can play a significant role in the betterment of individuals, groups and society at large.

Need for Value Education

Correct identification of our aspirations

The subject which enables us to understand 'what is valuable' for human happiness is called `value education' (VE). Thus, VE enables us to understand our needs and visualize our goals correctly and also indicate the direction for their fulfillment. It also helps to remove our confusions and contradictions, and bring harmony at all levels.

Understanding universal human values to fulfill our aspirations in continuity

Values form the basis for all our thoughts, behaviours and action. Once we know what is valuable to us, these values become the basis, the anchor for our actions. We also need to understand the universality of various human values, because only then we can have a definite and common program for value education and assured of a happy harmonious human society.

Fulfilling our aspirations

Values and skills are complimentary. To fulfill our aspirations both values and skills are necessary. When we identify and set the right goals, decide the right directions, this becomes a value domain, when we practice the values to achieve the goals, we develop the techniques to realize the goals by human endeavour, this becomes domain of skills.

For example, you want to lead a healthy life. Only wishing for good health will not help you keep your body fit and healthy. Without having understood the meaning of health, you will not be able to choose things correctly to keep your body fit and healthy.

Content of value education

The scope of value educational includes all dimensions (thoughts, behaviour, work and realization) and all levels (individuals, family, society and nature-existence).

The content of value education will include to understand yourself, your aspirations, your happiness; understand the goal of human life comprehensively, understand the other entities in nature, the innate-connectedness, the co-existence in the nature-existence and finally there role of human being in this nature/existence entirely.

Basic Guidelines for Value Educational

The following guidelines for the content of the course are important:

Universal

It needs to be applicable to all the human beings irrespective of caste, creed, nationalities, religion, etc, for all times and regions

Rational

It has to appeal to human reasoning. It has to be amenable to reasoning and not based on dogmas or blind beliefs.

Natural and verifiable

It has to be naturally acceptable to the human being, experientially verifiable, and not based on dogmas, beliefs or assumptions.

All encompassing

Value education is aimed at transforming our consciousness and living. Hence, it needs to cover all the dimensions (thought, behaviour, work and realization) and levels (individuals, family, society, nature and existence) of human life and profession.

Leading to harmony

The value education ultimately is targeted to promote harmony within the individual, among human beings and with nature.

Process of Value Education

The process for value education has to be that of self-exploration, and not of giving sermons of preaching the do's and don'ts. Whatever is found as truth or reality may be stated as a proposal and everyone is to be encouraged to verify it on his/her own right. Various aspects of reality facilitating of human values will be presented as proposals. We need to verify these proposals for our self and examine our living in this light.

In other words, "the process of value education is through self exploration and includes two things" verification at the level of natural acceptance, and experiential validation in living.

Thus value education is required to correctly identify our basic aspirations, understand the values which enable us to fulfill our basic aspirations ensure the complementariness of values and skills, and to properly evaluate our beliefs. It also facilitates the development of appropriate technology and its right utilization for human welfare.

17.4. ETHICS MEANING AND DEFINITION

The term ethics is derived from the Greek word *ethos* which can mean customs, habits, character or dispositions. At its simplest, ethics is a system of moral principle. They affect how people make decisions and lead their lives.

- Ethics is the word that refers to morals, values, and beliefs of the individuals, family or the society.
- Ethics is concerned with what is good for individuals and society and is also described as moral philosophy.
- Ethic is the discipline dealing with what is good and bad and with moral duty and obligation.
- Ethics refers to particulars set of beliefs, attitudes, and habits of individuals or family or groups concerned with morals. It is also used to mean `morally correct'

Reasons for studying ethics

Study of Ethics is very important because of following reasons

- To know the people's beliefs, values, and morals
- Learn the good and bad.
- Practice them to maximize well-being and happiness

Work Ethics

Work ethics is defined as a *set of attitudes concerned with the value of work, which forms the Motivational Orientation.* The 'work ethics' is aimed at ensuring the economy (get job, create wealth, earn, salary), productivity (wealth, profit), safety (in workplace), health and hygiene (working conditions), privacy (raise family), security (Permanence against contractual, pension, and retirement benefits), cultural and social development (leisure, hobby, and happiness), welfare (social work), environment (anti-pollution activities), and offer opportunities for all, according to their abilities, but without discrimination.

17.5 MORALS

Morals are the welfare principles enunciated by the wise people, based on their experience and wisdom.

They were edited, changed or modified or evolved to suit the geography of the region, rulers (dynasty), and in accordance with development of knowledge in science and technology, and with time.

Morality is concerned with principles and practices of morals such as:

a. What ought or ought not to be done in a given situations?

b. What is right or wrong about the handling of a situation?

c. What is good or bad about the people, policies, and ideals involved?

Difference between Morality and ethics

Morality is different from Ethics in the following ways:

Morality	Ethics
More general prescriptive based on customs and traditions	Specific and descriptive. It is a critical reflection on morals.
More concerned with the results of wrong action, when done.	More concerned with the result of a right action, when not done.
Thrust is on judgment and punishment, in the name of God of by laws.	Thrust is on influence, education, training through codes, guidelines, and correction
In case of conflict between the two, Morality is given top priority, because the damage is more. It is more common and basic.	Less serious, hence second priority only. Less common. But relevant today, because of complex interactions in the modern society.
Complex Character flaw, corruption, extortion, and crime.	Example: Notions or beliefs about manners, tastes, customs, and towards laws

Integrity

Integrity is defined as the unity of thought, word and deed (honesty) and open mindedness. It includes the capacity to communicate the factual information so that others can make well-informed decision. It yields the person's `peace of mind', and hence adds strength and consistency in character, decisions, and actions.

This paves way to success. It is one of the self-direction virtues. It helps to enthuse people not only to execute a job well, but to achieve excellence in performance. It helps them to own the responsibility and earn self-respect and recognition by doing the job.

Moral integrity is defined as a virtue, which reflects a consistency of one's attitudes, emotions, and conduct.

We do not act rightly because we have virtues or excellence, but we have those because we act rightly: Aristotl.

18

Goal, Mission and Vision of Life

GOAL – MEANING AND DEFINITION

Goals are an important part of getting what you want out of life. To put the concept very simply. Goals are the "how" you achieve something. It is roughly similar to purpose or aim, the anticipated result which guides reaction, to achieve an objective.

A goal is defined as a desired result or possible outcome that a person or a system envisions, plant and commits to achieve: a personal desire or end-point in some sort of assumed development. Many people endeavour to reach within a finite time by setting deadlines.

A goal is an observable and measurable, and result having one or more objectives to be achieved, within a more or less fixed timeframe.

Need to set personal goal

Individuals can set personal goals. A student may set a goal of a high grade in an exam. An athlete might run five miles a day. A traveler might try to reach a destination-city within three hours. Financial goals are a common example, like saving for retirement or to saving for a purchase.

Setting up of goals in life, planning and pursuing those leads to:

1. Specific effort aiming towards desired objectives of personal life. Fruitful and rewarding returns in all areas of personal life

2. Clarity in what to achieve helps in improving the efforts with full devotion and deep concentration.
3. Promotes long-term vision and short-term motivation.
4. Focusing on the intentions and desire.
5. Acquisition of related knowledge, and organisation of required resources.
6. Resolution of all right, inner conflict or limiting believes.
7. Unhindered progress and pride of accomplishment.

Goal setting

Goal-setting comprises a major component of personal development and management. Goal setting may involve establishing *specific, measurable, achievable, relevant*, and *time-bounded* (SMART) objectives. Goals can be long-term, intermediate, or short-term. The primary difference is the time required to achieve them.

Goal setting can serve as an effective tool for making progress when it ensures that group member have a clear awareness of what each person must do to achieve a shared objective.

Life is designed in such a way that we tend to look long-term but we live short-term. We dream for the future and live in the present. Unfortunately, the present can produce many difficult obstacles. But setting goals provides long-term vision in our lives. We all need powerful, long-range goals to help us get past those short-term obstacles. Fortunately, the more powerful our goals are, the more we will be able to act on and be confident of achieving them.

How to set the goals

1. Evaluate and reflect

Evaluation gives you a baseline to work from. First, take some time to think through and write down your current situation; then ask this question on each key point: *Is that OK*? The way we can reasonably decide what we want in future.

The purpose of evaluation is twofold.

i. It gives you an objective way to look at your accomplishments and your pursuit of the vision you have for life.
ii. It shows you where you are, so you can determine where you need to go.

2. *Define your dreams and goals*

One of the amazing things we posses as humans is the unquenchable desire to have dreams of a better life and the ability to establish and set goals to live out those dreams.

What are your dreams and goals? This is not what you already have or what you have done, but what you want. Have you ever really sat down and thought through your life values and decided what you really want? Have you ever tried to truly reflect, to listen quietly to your heart, to see move your heart into a within you? Your dreams are there.

Think about what really thrills your. When you are quiet, think about those things that really get your blood moving. What would you love to do, either for fun or for a living? What would you love to accomplish? What would you try if you were guaranteed to succeed? What big thoughts move your heart into a state of excitement and joy?

When you answer these questions you will feel great and you will be in the ``dream zone." It is only when we get to this point that we experience what our dreams are. Now, prioritize those dreams. Which are most important? Which are most feasible? Which would you love to do the most?

Put them in the order in which you will actually try to attain them. Remember, we are always moving toward action-not just dreaming.

3. Make you goals S.M.A.R.T.

The acronym S.M.A.R.T means Specific, Measurable, Attainable, Realistic and Time-sensitive.

Specific

Goals are no place to be vague. Ambiguous goals produce ambiguous results. Incomplete goals produce incomplete futures.

Measurable

Always set goals that are measurable. They should be "specifically measurable" to meet our principle of being specific.

Attainable

One of the detrimental things that many people to with good intentions is setting goals that are so high that they are unattainable.

Realistic

The root word of realistic is "real." A goal has to be something that we can reasonably make "real" or a "reality" in our lives. There are some goals theat are simply not realistic. You have to be able to say, even if it is a very difficult goal, it is entirely realistic that you could make it.

Time

Every goal should have a timeframe attached to it. One of the powerful aspects of a great goal is that it has an end a time in which you plan to accomplish it. You must work on it such that you do not want to get behind; you work diligently because you want to meet the deadline.

4. Have accountability

When you decide and declare your goals are, you are going to be held accountable. You need to explain where you are in the process of achieving that goal. Being accountable further drives you to give your best efforts and achieve the goals in their totality, within the desired time limit.

Mission and Vision of Life

The word "mission" originates from 1598 when the Jesuits sent members abroad, derived from the *Latin missionem* (nom. Mission), meaning "act of sending" or *mittere*, meaning "to send"

A **Mission** Statement defines the objectives of life and approach to reach those objectives. Elements of **Mission and Vision** Statements are often combined to provide a statement of the purposes, goals and values in the life.

A personal mission statement is a powerful tool because it provides you with a path a success, and it gives you permission to say no to the things that are distractions. It also changes over things "As we get older, we have more life experiences and acquires new skills,"

Bart estimates that in practice, only about ten percent of mission statement say something meaningful. For this reason, they are widely regarded with contempt.

How to write your own personal mission statement

While you write a personal mission statement for yourself, there is power is sharing it. "The more you share, the more support you get to achieve your mission", "Friends and members can support you or call you out if you are doing something counterproductive."

Steps for writing personal mission

1. Examine the lives of others

Think of a person in history or in your life whom you admire. What are the qualities of that person that you would like to emulate. These qualities can relate to their character, values, achievements, personality, or simply the way they live their lives. Consider the specific reasons you admire the person, and list those qualities in detail.

2. Determine your ideal self

Define the type of person you event to become, not just what you want to have or achieve. This ideal should reflect your core values and your definition of living with integrity. Consider all areas of your life, as a spouse, friend, employee, parent, etc. and who you went to be in each of those roles.

This exercise may take some time, but it is well worth, the effort beyond its usefulness for your mission statement. It helps you to clarify your personal operating system and reminds you of what you are capable of becoming.

3. Consider your legacy

Determine all of your life roles (career, family, community, etc.) and write down a short statement of how you would like to be described in each those else. Think about how you would like the important people in your life to remember you and talk about you.

For example, you might want your boss to say, "He was a man of character and integrity who was a compassionate and inspiring leader and a visionary for our organisation."

This exercise may feel awkward, but no-one else has to see it. It is to help you decide how you want to step into each of the roles in your life and to clarify in concise words how you want others to perceive you.

4. Determine a purpose

Write down a purpose for the four fundamental elements of who you are: physical, mental, emotional and spiritual. In each of these areas, what is the most important way you want to express yourself? This may seem confusing, but it is really simple. We tend to spend our lives without considering our purpose for our essential being and how we want to achieve that purpose.

5. Clarify your aptitudes

What are the talents and skills you possess that are most important to you and that you actually enjoy? Part of your mission statement should reflect you best aptitudes and strengths: these are what create joy and energy in our lives. When we spend our time on what we do well and enjoy, our lives not only have meaning also have vibrancy.

Make a list of all of your personal and professional talents, aptitudes, and skills, even those you may take for granted, like being a good friend or having the ability to organize well. Then circle the skills you enjoy or find fulfilling. Focus on these.

6. Define specific goals

Base on everything you outlined above, what are some related you want to achieve in your life? Goals are specific, measurable, achievable, results-focused, and time bound.

As you look at the people who inspired you, the ideal self you've defined, the legacy you want to leave, the purpose you have for the elements of your life, and the aptitudes you want to enjoy, what are the outcomes you want to achieve for your life?

7. Craft your mission statement

Taking into account steps 1–6, being crafting your personal mission statement. Keep it simple, clear and relatively brief from a few sentences to a couple of paragraphs. You can write it as a statement that flows or with bullet points. How you craft it is less important than what you want to express.

The point is you want a statement that will guide you in your day to day actions and decisions, as well as your long term goals. Try to keep your words positive and affirmative. Focus on what you want rather than what you do not want.

Vision of Life

The word vision comes from the *Latin visionem,* meaning wisdom. Having a vision in having the wisdom to see past certain events in expectation of a greater outcome. All visions become reality. as a result of three factors: creativity, perseverance, and actions. Vision, on the other hand, is the "why". A Vision is a picture or ideas you have in your mind of yourself, your business, or anything that is going to happen

Need for having a Vision for Life

A clear vision helps you pursue dreams and achieve goals:, and idea of the future, a strong wish.

1. You feel much more valuable as a person when you set and achieve vision and goals. In fact, a well-defined personal vision should naturally leads to more focused goals set.
2. Our personal vision begins to manifest once we learn how to use our three eyes: Our two eyes may see the new house we're thinking of buying, for example, but our mind's eye sees a home for many years, a growing family, and the chance for a great investment.

3. Our personal vision statement guides our life and provides the direction necessary to define the course of action and choices you make about your career. Your personal vision statement is the light shining in the darkness toward which you turn to find you way. It illuminates your way.

A career vision is a statement about what you want to achieve in your **career** – the major accomplishments you hope to attain, the level or position you hope to rise to, and the lasting impacts you hope to make. **A career vision** statement is very much "big picture" of your thinking.

How to Create Your Life's Vision

1. Identify what matters in life

What is the real meaning of life? How should you live your life? What things you want to do and what recourses will be needed to achieve them?

2. Make a list of the categories of things that matter to you

It's all about what matters to you, and what you want out of your brief time on this planet.

- Health – exercise, diet, mindfulness, perspective
- Ability – skills, knowledge, character
- Relationship – curate and cultivate them
- Time – wisely using whatever time you have
- Wealth – creating the value necessary to support goals
- Experiences
- Accomplishments
- Contentment – being happy with who you are is perhaps the ultimate goal

Now, for each of your categories, write down what you want or need from each. Think about the things you want to accomplish. Finally, craft a statement that describes what your ideal life looks like. Your vision statement will consist of an overall description of your life, combined with a list of areas that matter most, and high-level goals for each area.

3. Build a system for yourself

Review your vision and goals regularly, and update your action plan for accomplishing those goals.

Your main priority should be making your system. Start with calendar reminders and task list items and build life planning time into your daily and weekly routines until it becomes habit.

George Bernard Shaw said: "Life isn't about finding you. Life is about creating your".

The majority of us take life as it comes. More focused on approaching events and current situations, we often do not think in the long term. And that is understandable, because the present affects us now. But broadening our vision leads to totality of being. Think of cooking a meal. Before you even begin, you have an understating of your final result. You know what you are making, what ingredients to add, in what order, and the cooking time. If you were to cook without knowing the steps, you would end up with anything but a good meal. Similarly, if you try to create the life you want without knowing what steps to take, you will live constantly short of your dreams.

4. Create a bigger version of yourself

We have the tendency to place ourselves within boundaries that define who we can and cannot be. This mindset causes us to settle for the lowest level of our potential. Painting a mental image helps you understand what it is you really want.

5. Write down your objectives

We all have ambitions, and some are more significant than others. Make a list of your ten most important aspirations and circle at least one you plant to achieve within the next year.

6. Plan a strategy and timeline

A strategy specifies the actions you will take in order to advance towards a goal. If your vision is to have your own home, your strategy should include how to earn or save the money to buy one.

Be realistic about timing and consider all factors that could work for or against you. Keeping a timeline also which keeps you on track with your intentions.

7. Check your emotions

Our vision fogs up when emotions are involved. We simply cannot visualize the outcome of something to which we have become attached. When you hit a wall of traffic, you become stuck and cannot see up the road to detect what is causing the jam. Your vision is blocked. This is what happens when your own emotions create a wall before your mind's eye, keeping you from anticipation what is to come in any given situations. By learning to control your emotions you can detached long enough to see the truth of what awaits.

8. Take the first step

Initiate action according to your intentions. Your first step may be a leap of faith, but it is with this first step that you prove to yourself your own abilities.

9. Envision your outcome along the way

Keep your eye on the goal. Don't allow yourself to become distracted or deterred from your goals. Continue to practice the same vision you have always dreamed to be your reality, the one you know you deserve.

10. See the bigger reason behind setbacks

Setbacks make you cautious more focused and you are bound to try harder. Always remind yourself that there is a greater purpose to all events.

11. Ask for help

The universe wants you to succeed and will guide you if you just ask. Help may not come immediately but when it does, you will receive more than you imagined and from unexpected sources.

12. Pay attention to the people you meet along the way

Everyone who comes into your life serves a role and a purpose. Encounters with negative minded people and negativity should only reassure you of your own strengths.

13. Never, ever give up

Simply having a vision does not fulfill it; you must work to materialize your dream. You will experience bumps along the road and come across people who will try to keep you from succeeding, but you must not allow yourself to be intimidate. Remember that you are on a mission to achieve something incredible, and it will take a lot to stop you.

14. Celebrate the achievements

Your journey must include time of reflection. Compare where you were five years ago with where you are now and congratulate yourself for your achievements both big and small. Do not forget to reward yourself for your hard work, careful planning, and strong character.

15. Never repeat old mistakes

Learning entails personal evolution, evolving to take new actions and to stay away from past actions. Learn from your mistakes and do not put yourself through situations which you know will attract a similar outcome.

■■■■

19

Principles and Philosophy of Life

PRINCIPLES

Meaning and Definition

A principle is something primary that helps in explaining phenomena. A principle can be some existing factor in nature **(principles of nature and being)**, or it can be a logical proposition or judgment *(principle of reason)* that is a starting point of a valid argumentation.

- Principles are defined as the fundamental norms, or values that represent what is desirable and positive for a person, group, organisation, or community, and help it in determining the rightfulness are wrongfulness of its actions. Principles are more basic than policy and objectives, and are meant to govern both.
- Principles are the fundamental truth or propositions that serve as the foundation for a system of belief or behaviour or a chain of reasoning.
- A principle is a basic idea or rule that explains or controls how something happens or works.
- A principle is a concept or value that is a guide for behaviour or evaluation. It is a substantive and collective term referring to rule of governance, the absence of which, being "unprincipled"

Principles of Life

1. No Strain, No Gain

The saying no strain, no gain in 100 per cent true. If you do not strain your muscles, they get weak. If you do not strain your brain, your mental ability will decline. If you do not test your character, you become spineless.

We often talk about doing all kinds of crazy things – climb mountains, run marathons, skydive, start businesses, travel the world, write books, record albums, make movies, the list is endless. Just think for a second what exactly you want to do. Now just think why you have not done those things. Do not shy away from hard things. Instead, train yourself to become a person that can endure difficulties "*if you rest, you rust.*"- *Helen Hayes*

2. Avoid Negativity at every cost

Do not put negativity at work at, home, with their friends, family. In fact, people are inherently negative. That is why see so much complaining, lying, blaming, jealously, bullying, in the world.

All the negative thoughts keep you from living a healthy and wealthy life. So why do you let yourself be exposed to negativity.

3. Give More than You Take

Helping others is the most important thing you will ever do in your life. Habit of taking things tends to stick with us as we grow up. We think it is normal to take, take, and take. In fact, we think we should get everything we want. Life is not about taking. So start giving more. Enjoy the joy of giving.

"The effect you have on others is the most valuable currency there is. "- Jim Carrey.

4. Time is more valuable than Money

Out of all the resources in the work, time is the most valuable resource you have. Be more selective with your time. Do not waste it like you can always get it back. Sure, you can always earn back the money that you have spent, but you can never earn back time. Once you spend time, it is lost forever.

5. Create Your Own Path

We look at others and view them as examples: Fathers, mother, brothers, sisters, friends, bosses, mentors, authors, entrepreneurs, artists. Learning from others is one of the best things to do. But remember you are not all those people. You are you. That means one simple thing, *Get off the beaten path, and create your own way.* Of course, travelling the road that has been paved by others is much easier. But it is not always rewarding. Accomplishment weights more than money, status, and even comfort.

"Men almost always walk in paths beaten by others and act by imitation" - Niccolo Machiavelli

6. Do Whatever Life expects of you

Life is random. Instead of wishing that things were different, accept the circumstances of your life. No matter how bad they are. Things are what they are. Look at it his way, "Regardless of where you are in life, you are there for some reason may be God, the universe, or invisible forces".

7. Life Move In one Direction

You can live in three different time dimensions in your head. The past, the future and the present. If you live in the past, you are stuck in "you" mode. That means you always think about why things happened. That is a recipe for unhappiness. If you live in the future, you live in "what if" mode. That state of mind causes anxiety about everything that might happen in the future. Again, that is a recipe for living a timid life. You must know very well that there is only one culture time dimension, and that is present.

PHILOSOPHY

Meaning and Definition

The original meaning of the word philosophy comes from the Greek roots philo-**meaning** "love" and –sophos, or "wisdom" When someone studies **philosophy** they want to understand how and why people do certain things and how to live a good life. In other words, they want to know the meaning of life.

The term **philosophy** encompasses the study of general and fundamental problems concerning matters such as existence, knowledge, values, reason, mind, and language.

Philosophy is the study of the fundamental nature of knowledge, reality, and existence, especially when considered as an academic discipline, a theory or attitude that acts as a guideline principle for behaviour.

Philosophy is the study of wisdom, understood as the ability to conduct the human activities; and also as the perfect knowledge of all the things that a man can know for the directions of his life, maintenance of his health, and knowledge of the arts.

Philosophy is a study of problem which are ultimate, abstract and very general. These problems are concerned with the nature of existence, knowledge, morality, reason and human purpose.

Definition of philosophy of life

Philosophy is defined as *an overall vision of attitude toward life and the purpose of life.*

Philosophy is also defined as, the most basic beliefs, concepts, and attitudes of an individual or group. The focus is on *personal philosophy*, and the essential philosophical elements are centred on belief, concepts or ideas, and attitudes.

Branches of Philosophy

Metaphysics	Study of Existence	What's out there?
Epistemology	Study of Knowledge	How do I know about It?
Ethics	Study of Action	What should I do?
Politics	Study of Force	What actions are permissible?
Aesthetics	Study of Art	What can life be like?

Importance of Philosophy

A philosophy is a system of beliefs about reality. It is one's integrated view of the world. It includes an understanding of the nature of existence, man, and his role in the world. Philosophy is the foundation of knowledge. It is the standard by which ideas are integrated and understood.

1. Philosophy is a necessary product of man's rational mind.
2. To live, man must gain knowledge of the world. To understand the world, man must form conclusions about its very nature. For instance, to gain knowledge of particular objects, man must recognize that objects have identity. He must recognize that conclusions are possible because the world does exist, and in a particular way.
3. Philosophy provides the framework for which man can understand the world.
4. It provides the premises by which man can discover truth, and can use his mind to support this life. Every man has an understanding of the world. Every man must have philosophy, even if it is never made explicit.
5. When you love wisdom, when you tend to use your mind more, then you will be able to make better decisions, rational decisions, rather than decisions based on fear, anger or pain.

Wisdom will enable you to see through the many false things in our world, such as materialism, sexism, and a whole host of other lies. Based on these points, philosophy will even help you in the area occupational therapy, because it will teach you to work hard and to be fair.

20

Self Exploration

MEANING AND DEFINITION

In order to know where you want to go in life, you have to know who you are. Self exploration is the process of finding out what is valuable to you by investigating yourself. What is right for you is to be judged by yourself only since it is only you who can feel happy, successful, unhappy or unsuccessful. Therefore, you are the best judge to decide what is right for you. This whole process is called self exploration.

Exploration means Observing outside and Self-Exploration means Observing Inside

The examination and analysis of one's own unrealized spiritual or intellectual capacities is self exploration.

Following are some definitions of self exploration

1. According to Merriam-Webster: The examination and analysis of one's own unrealized spiritual or intellectual capacities is self exploration. It is a process to find out what is valuable to me by investing within me.
2. According to Ryan Howe's, a psychologist, writer and professor in California, self-exploration involves "taking a look at your own thoughts, feelings, behaviours and motivations and asking why. We are answers to all the questions we have about ourselves".
3. Self exploration is just finding answers to the following fundamental questions of all human things:

a. **What are my Desires/Goals?**

b. **What do I really want in life, or what is the goal of human life?**

c. **What is my program for fulfilling the desire? How to fulfill It?**

d. What is the program to actualize the above?

4. It is a process of self evaluation through self investigation. It is a process of dialogue between "what you are, what you really want to be".
5. Process of knowing human conduct, human character. Process of knowing innateness. Process of being in harmony with oneself and with entire existence.
6. It can be attempted by being the keen observer, Focusing on solutions rather than on problem, Chasing strong emotions, Identifying core values and beliefs and Knowing line of demarcation and responsibility.
7. Self exploration takes place in the self and not in the body. It may be regarded as a philosophy and methodology in which the direct experience and focused reflections of the individual helps to increase knowledge, develop skills and clarify values. Experimental validation is a process that infuses direct experience with the learning environment and content.
8. It is a process of identifying our innateness (Svata) and moving towards self-organisation (Swantantrata) and self-expression (Swarajya)
9. Self exploration is finding the answer to the desires (Aspirations) and the plans to fulfil these desires.

The purpose of self exploration is as follows

1. Self esteem

It helps people understand and accept who they are and why they do, what they do, which improves self-esteem, communication and relationship.

2. Realization and understanding

The process of Self-Exploration results in `Realization' and `Understanding' in us. This realization and understanding leads to the following:

- Assuring
- Satisfying

3. Strengthening intention and will

The stronger are you, intention the more you will proceed in alignment with your desire. If your intention of will to proceed is ambiguous or small, then the will and intention of those who present the practices or techniques you use, will take precedence your own.

4. Developing faith in yourself

Faith in yourself makes you to keep going. Develop faith in your responsible to achieve everything or if that is too much then you must have faith in the possibility to develop the abilities to achieve everything.

5. Self responsibility

Self responsibility helps in understanding that only you are responsible for yourself, and only you are responsible for what you achieve within your life. No one else is to be blamed for what you choose to manifest and experience and no one else it to be blamed for what you choose to not manifest and experiences. When you pass the responsibility of finding yourself to others then how will that give you the ability to be responsive to yourself?

6. Self honesty

Self honesty is a very important attribute and should be either cultivated or enhanced. Self honesty requires that you make changes if you are not moving closer to your goal. Self honesty means that you cannot hide behind yourself and put the blame of stagnation on others or on factors out there. Self honesty means that you will be honest about all parts of yourself, whether you like these parts or not.

Self honesty means that you will always try and look at the illusions you are presenting.

7. Courage

For most people, this is a very difficult area to master. It takes much courage to put forward everything that you think, you know and to stand for it:

8. Dedication and perseverance

These values ensure that when you experiences the fear of the unknown and the hardships of changes, you keep going till you achieve the desired goals.

9. Humility and humor

Humility is not always claiming to know best which helps you accept when you make mistakes, and humor helps you so that you neither take yourself nor your path too seriously. Both these traits help in performing the tasks smoothly without having an ego and always yield very good results.

■■■■

21

Self Awareness

MEANING AND DEFINITION

Self awareness is the first step in creating what you want and mastering in your life. The focus of your attention, your emotions, reactions, personality and behaviour, determines where you go in life.

Self Awareness is having a clear perception of your personality, including strengths, weakness, thoughts, beliefs, motivations, and emotions. Self Awareness allows you to understand other people, how they perceive you, your attitude and your response. Self-awareness is really just about being aware and confident of who you are. It can relate to knowing your own values, beliefs, personal preferences and tendencies

Having self awareness allows you to see where your thoughts and emotions are taking you. It also allows you to take control of your emotions, behaviour, and personality so you can make the changes you want. Until you are aware of your thoughts, emotions, and behaviour, you will have difficulty in making changes in the directions of your life.

Self-awareness is partly knowledge about the self: strengths and weaknesses, opportunities and threats. It can be derived in many ways from many sources. Sometimes self-insight comes from a sudden relation in the classroom or on the couch. It can even occur during a work appraisal. It comes out of both experiences of success and failure. It may come from what others say or by receiving feedback from a personality test.

Self-Awareness is defined as the accurate appraisal and understanding of your abilities and preferences, their implication for your behaviour and their impact on others. It's reality-testing; a calibration against the facts of life.

Self-awareness is the first step in the career development process. When making career decisions, It is important to the four areas of 'yourself concept':

1. Personality
2. Interests
3. Skills
4. Values

Increasing your awareness of these four areas will help you to choose a career path that will be a better "fit" for you. Salary means nothing if going to work every day makes you miserable! There are many great careers out which are consistent with your personality, skills, values and interests. To get an idea of what they might be, meet with a career counselor.

Why to Develop Self Awareness?

1. Self awareness is important because when you have a better understanding of yourself, you are able to experience yourself as a unique and separate individual
2. You are able to make changes in the thoughts and you make up your mind.
3. Changing the interpretations in your mind allows you to change your emotions.
4. When you can change the interpretation in your mind of what you think, you can change your emotions and shift the emotional quality of your relationships in achieving the success
5. Self awareness is one of the attributes of Emotional Intelligence
6. Self-Awareness helps to make changes and to build upon areas of strength as well as indentify the areas where you would like to make improvements.

7. Self-Awareness helps to tune the feelings, as well as the behaviour and feelings for others.
8. By becoming self-aware and understanding your strength and limitation, you open up opportunities the just are not available otherwise.
9. You can also able to have more honest and genuine relationships because the people that you are attracted to, will be attracted toward you.

The ways to become self-aware

Developing your self-awareness helps you learn more about yourself and what you are capable of. There are some great ways can work on your self-awareness

1. Assess your self-talk

The first step in self-awareness is to listen to you. What is going on in your mind? Is it a series of negative thoughts that make you feel helpless or inferior to others? Or are you always looking on the brighter side?

In practice: Take a couple of minutes each day to sit in silence and listen to the tone of your inner voice. One way of getting your inner voice going, is to stand in front of mirror and hear what you are saying to yourself about how you look. It might even help to write down your thoughts so that you can get a better idea of how positive or negative they are.

2. Use your senses

Your senses (sight and sound, in particular) can provide you with huge insights into your own and other people's feelings, and situations in general. But these senses are often viewed through

The filter of our self-talk. For example, a frown doesn't always mean that someone is angry, and a groan does not necessarily means that the person you are talking to is bored, despite what your inner voice might be saying.

In practice: The next time you feel that someone is judging you, or has made you feel bad about yourself, take a step back and write down why you think so. Ask yourself, `Could I have interpreted what was said/done differently?' You might find that your interpretation was clouded by your own negative thoughts.

3. Tune into your feelings

This can be tough if you are not the kind of person who likes to think too deeply about your feelings. Your feelings are spontaneous and emotional responses, to the whatever you experience. Like your senses, they give you good information about what is going on around you, should you choose to tune into them.

There are some physical signs that you can look for that might help you to `read' your feelings They include:

- A warm feeling in your face might mean you are embarrassed.
- A feeling of 'butterflies' in your tummy can means you are nervous.
- Clenching your teeth might mean you are angry.

In practice: Be aware of physical signs that might indicate how you are feelings. By engaging with how you are feeling, you can get better insights into what you like, what makes you feel uncomfortable and what makes you angry.

■■■■

22

Self-Satisfaction: The Ultimate Goal of Life

MEANING AND DEFINITION

Almost everyone, at some point or the other in their life, might have asked themselves about the reason of their birth, the purpose of their life, and the goals that they would like to pursue. These questions, if left unanswered continuously keep posing troubles to the individual and lead to an unknown displeasure and unhappiness.

No matter how much success a person gets, no matter how rich a person becomes, and no matter how many goals are achieved, somehow these questions keep on confronting the person. This will continue until and unless such accomplishments do not lead to self-satisfaction.

The only answer to these questions is to seek self-satisfaction. It is self-satisfaction that can keep an individual away from these troubling questions. Self-satisfaction should thus be the sole quest or search, for human beings. Self-satisfaction should therefore be the ultimate goal of life.

As simple as may it sounds, achieving self-satisfaction is not at all an easy task. But, there are ways in which one can constantly look out for self-satisfaction and in the process find answers to the many unanswered questions that life imposes upon us.

Definition

- Self satisfaction is defined as the pride worthy satisfaction with oneself or one's position of achievements.
- The feeling you have when are satisfied with yourself is called self satisfaction.
- Self-satisfaction is a sense of absolute fulfillment with oneself and one's own accomplishments.

Finding meaning in life for self-satisfaction

Frankl suggests three ways of finding meaning in life.

1. By Experiencing something or someone valued by the individual. This mainly includes experiencing the value of other persons, such as parents, siblings, friends, lover, etc. In order to develop meaning in the life of the valued person, the individual can develop meaning in his own life.
2. By involving oneself in creative deeds, such as art, music, writing, invention, etc.
3. By developing virtues such as compassion, bravery, good sense of humor, etc.

Once a person finds meaning in life, he sets off to a journey of achieving self-satisfaction.

Carl Rogers says that people have a real self and an **ideal self.** The real self in what the person actually is or can be. It is true potential of the person. The **ideal self** is what the person feels that he should be which may not be the real potential of the person.

People, at times, tend to move towards their ideal self rather than their real self, due to expectations or external pressure. This creates a gap between real self and ideal self. Rogers calls this gap as **incongruence**. If there is no gap between a person's real self and ideal self then it is referred to as **congruence.** The more the gap between the real self and the ideal self, the more there is inner conflict and displeasure. The individual feels threatened and anxious. They become defensive, and their thinking becomes rigid and constricted. An individual is well-adjusted and mentally healthy in a state of congruence, while in the state of incongruence the person is maladjusted.

According to Rogers, we should try not to conform to the expectations of others. We should rather come to terms with our own nature, trust one's own experience, and accept the fact that other people are different. In this way, the gap between the real self and the ideal self will become lesser and a state of congruence will come into place.

A high level of self-awareness is required for self-satisfaction. An individual should have a realization of own strengths and weakness. This will make the person know how good or bad he is at various things. This wills also the person more focused and will enable him to know exactly what is desired from life. When a person knows exactly what he wants to do, it generates an immense amount of positive feeling within that person and leads to a lot of self-worth.

Ways of enhancing life satisfactions:

1. **Friends** – Having a lot of close friends, can boost life satisfaction are nearby 20 per cent.
2. **Have Goals** – People with goals are nearby 20 per cent more satisfied with their lives. People who could identify a goal they were pursuing were 19 per cent more likely to feel satisfied with their lives and 26 per cent more likely to feel positive about them (Krueger, 1998).
3. **Have a Life Story** – Research shows that meaning in life comes from the stories we tell ourselves about our lives.
4. **Money does not matter** – The more materialistic people are, the less satisfied they are with their lives. Among participants in one study, those with most materialistic values, rated their lives as the least satisfying (Ryan and Dziurawiec, 2001).
5. **Changes in behaviour** – People over forty who could identify at least one change in their viewpoints or behaviour in recent months were and percent more likely to feel hopeful about the future and 5 per cent more likely to say they were generally in a good mood (Gross Baum and

Bates, 2002). Older people who continue to read and learn are much more satisfied with their lives.

Ways to be Contented, Happy and Satisfied with Life

In order to be happy with what you have, you need to follow a few simple rules.

1. Be thankful for everything you have now

Be happy, whenever you are feeling down, think of something that makes you feel happy and calm. Enjoy happiness. Be grateful for everything, whether it is good on the table or a nice place to sleep at night or anything else. You need to take the time out of your day and give thanks either loudly or within your mind, for all the things you have. This simple task takes less than a minute and will have a dramatic impact on your current happiness.

2. Live in the moment

Stop being frustrated. Replace anger. Be calm. Relax and/or meditate. Stop comparing yourself with others. Forgot about the past or thinks that have not happened yet. Living in the moment will get you involved in what is going on right now. Take the time to explore your senses.

3. Follow your dreams and passions around

Do not worry about how you can make the most money or how you are going to get a new car like your neighbour has. Worry about what you need to follow in life to become happy. If you like animals get a cat. If you like nature goes for a walk. It is that simple.

4. Love yourself for who you are

Do not try to be someone you are not. You do not to change anything about yourself in order to by happy and content.

5. Strive for the best and settle for whatever you end up with

Do not worry if you don't come out on top. Most of us do not. Grieve it your best shot and be happy where you end up.

6. Start moving right now

It does not matter which direction you go, just pick one and start moving that way. If you want to lose weight then get up and start exercising right now.

7. Avoid anger and negative energy

Be positive and optimistic. Do not worry yourself with a negative attitude or by being around with negative people. Life is too short to be pessimistic mood. Overcome addictions.

Happiness does not depend on what you have what your are, it sole relies on what you think – Buddha.

■■■■

23

Decision Making

MEANING AND DEFINITION

The term decision making means the process of deciding about something important, especially in a group of people or in an organisation - *The oxford Advanced Learner's Dictionary*

- In its simplest sense, decision-making is the act of choosing between two or more courses of action. In the wider process of problem-solving, decision-making involves choosing between the possible solutions to problem. Decision can be made through either an intuitive (based on feelings) or a reasoned (based on logic) process, or a combination of the two.

- Decision making can be defined as the process of making choices by identifying a decisions, gathering information, and assessing alternative resolutions.

- Decision-making involves the selection of a action from among two or more possible alternatives in order to arrive at a solution for a given problem"- *Trewatha and Newport*

- Decision making involves the thought process of selecting a logical choice from the available options. When trying to make a good decision, a person must weigh the positives and negatives of each option, and consider all the alternatives.

- Decision can be defined as a course of action purposely chosen from a set of alternatives to achieve organisation or managerial objectives or goals.

Decision making in life

Our decisions determine the outcome of our lives. So if you want life to work out way you want it to, then make sure you keep making right decisions. Events in our life shape us and our choices define us.

People forget the great power they hold within that will shape life and they ignore the importance of decisions making.

Our life is designed by the choices we make. You are the who chooses how to live your life or create your life's path. So make the decisions which will take your life in the directions you want.

"Success and failure are not overnight experiences. It is the small decisions along the way that cause people to fail or succeed". *Anthony Robbins*

Sometimes we rush into making decisions and regret the choice we made. A hasty decision cloud lead to long term disappointment. So it is best to think through everything very clearly, but do not delay the decisions to long, because sometimes you will go past the point where your decisions matter.

Some tips for making decisions in life

1. Define the goal and objective of your life.
2. Take into account all important factors that matter like your family background, available resources, prevailing circumstances, values and ethics.
3. Based on the above, think of all possibilities/alternatives to reach to the goal of your life.
4. Consult your friends/family members and obtain their suggestions.
5. Weigh all the alternatives, accord them priority and choose the best one.
6. Take your own decision.
7. Learn to take timely decision, but do not take hasty decision.

8. Try out the decision on a pilot basis.
9. Review the outcomes.
10. Repeat the same process and verse the decisions based on the outcomes.

Decision Making in Organisation

Decision making is very important both in personal life as well as for the organisation. Decision-making is an integral part of modern management. Decision making process can be regarded as check and balance system that keeps the organisation growing, both in vertical and linear directions. It means that decision making process seeks a goal.

The **goals** are pre-set business objectives, company missions and its vision. To achieve these goals, company may face a lot of obstacles in administrative front, marketing wings and operational domains. Such problems are sorted out through comprehensive decision making process. No decision comes as an end in itself, since in may evolve new problems to solve. When one problem is solved another arises and so on. Hence decision making, is a continuous and dynamic process.

Decision making is taken as primary function of management. Every manager takes hundreds and hundreds of decisions subconsciously or consciously making it the key component in the role of a manager. Decisions play important role as they determine both organisation and managerial activities. Decision making process is continuous and indispensable component of managing any organisation or business activity. Decisions are made to sustain the activities of all business and organisational functioning.

Decision is made at every level of management to ensure organisational or business goals are achieved. Further, the decisions make up one of core functional values that every organisation adopts and implements to ensure optimum growth and development in terms of services and/or products offered.

Decision making process is a consultative affair done by a committee of professionals to run any organisation in a better way. Since it is an ongoing activity, decision making process plays vital role in the functioning of an organisation. As intellectual minds are involved in the process of decision making, it requires solid scientific knowledge coupled with skills and experience in addition to mental maturity. A lot of time is consumed while decisions are to be taken. In a management setting, decision cannot be taken abruptly.

Process of Decision making

Since decision making process follows some sequential steps, a lot of time is spent in this process. This is the case with every decision taken to solve management and administrative problems in a business setting. Though the whole process is time consuming, the result of such process in a professional organisation is magnanimous.

Use of step-by-step decision-making process, by Organising relevant information and defining alternatives, can help in making more deliberate and thoughtful decisions. This approach increases the chances of choosing the most satisfying alternatives possible.

Some simple steps for decision making are

Step 1: Identify the decision

First of all, understand the problem. Try clearly to define the nature of the decision you want to take. This is very important step.

Step 2: Gather relevant information

Collect some pertinent information and data before you make your decision. Decide upon what information is needed, identify the best sources of information, and the ways to get it. This step involves both internal and external "work." Some information is internal: you'll seek it through a process of self-assessment. Other information is external: you'll find it online, in books, from other people, and from other sources.

Step 3: Identify the alternatives

As you collect information, you will probably identify several possible paths of actions, or alternatives. You can also use your imagination and additional information to construct new alternatives. In this step, you will list all possible and desirable alternatives.

Step 4: Weigh the evidence

Draw on your information and emotions to imagine what it would be like if you carried out each of the alternatives to the end. Evaluate whether the need identified in Step1, would be met or resolved through the use of each alternative. As you go through this difficult internal process, you will begin to favour certain alternatives. Those seem to have a higher potential for reaching your goal. Finally, place the alternative in a priority order, based upon your own value system.

Step 5: Choose among alternatives

Once you have weighed all the evidence you are ready to select the alternative that seems to be the next among those identified. You may even choose a combination of alternatives. Your choice in this step may very likely be the same or similar to the alternative you placed at a top of your list at the end of Step 4.

Step 6: Take action

You are now ready to take some positive action by beginning to implement the alternative you chose in Step 5.

Step 7: Review your decision and its consequences

In this final step, consider the results of your decision and evaluate whether or not it has resolved the need you identified in Step 1. If the decision has not met the identified need, you may have to repeat certain steps of the process to make a new decision. For example, you might want to gather more details or somewhat different.

■■■■

24

Motivation

MEANING AND DEFINITION

The word Motivation has been derived from the root motive which means the latent power in a person which impels him to do a work. Motivation means movement or motion, an inner state the energizes, activates, moves and directs human behaviour towards goals. It is a need satisfying and goal seeking behaviour.

- Motivation is the process initiating a conscious and powerful action. It explains why people do the things, they do.
- Motivation is a theoretical construct used to explain behaviour. It gives the reason for people's actions, desires, and needs.
- A motives is what prompts the person to act in a certain way, or at least develop an inclination for specific behaviour.
- Motivation is defined as the process that initiates, guides, and maintains goal-oriented behaviours.
- Motivation is what causes a person to act.
- Internal and external factors that stimulate desire and energy in people to be continually interested and committed to a job, role or subject or to make an effort to attain a goal, define motivation.

- **Motivation means a process of stimulation people to action to accomplish desired goals – W.G. Scott.**
- **Motivation is the result of interaction between both conscious and unconscious factors such as:**

1. **Intensity of desire or need**
2. **Incentive or reward value of the goal**
3. **Expectations of the individual and of his or her peers.**

These factors are the reasons for behaviour of anyone in a certain way.

Common misconception is that motivation is a personality trait, and that some people have this quality, while others do not. However the fact is that motivation depends on the individual as well as on the situation. Situation plays an important role in determining the level of motivation in a person at a given time. Every individual has different motivational drives in different situations.

Importance of Motivation

To be successful in life, motivation is a must. It is important in order to achieve

1. High Level of performance
2. Good Human Relations
3. Increase in moral
4. Proper use of human resources
5. Helpful in achieving the goals

Process of Motivation

Koonz and O'Donnel have shown the process of motivation as need gap satisfaction chain in the following way:

Need Desire Tension Action Satisfaction

Types of Motivation

It is important to understand that we are not all same. Thus motivating anybody effectively requires gaining an understanding of the different types of motivations. Such an understanding will enable better categorization of the team members and apply the appropriate type of motivation. Each member is different and the motivational needs of each member will also be varied.

There are two types of motivations.

1. Intrinsic Motivation

Intrinsic motivation means that the individual's motivational stimuli are coming from within. The individuals has the desire to perform a specific task, because its results are in accordance with his belief system or fulfils a desire and therefore importance is attached to it.

Our deep-rooted desires have the highest motivational power.

Below are some examples:

- **Acceptance**: Need to feel that we, as well as our decisions, are accepted by others.
- **Curiosity**: The desire to know.
- **Honour**: Need to respect the rules and to be ethical.
- **Independence**: Need to feel we are unique.
- **Order**: Need to be organised.
- **Power**: Desire to be able to have influence.
- **Social contact**: Need to have some social interactions.
- **Social Status**: Desire to feel important.

2. Extrinsic Motivation

Extrinsic motivation is external in nature. Extrinsic motivation mans that the individual's motivational stimuli are coming from outside. In order words, our desire to perform a task are controlled by an outside source. Note that even though the stimuli are coming from outside, the result of performing the task will still be rewarding for the individual performing the task.

So, extrinsic motivation does not always have to be another person, but it is some outside demand, obligation, or reward that requires the achievement of a particular goal. Intrinsic motivation, however, is an internal form of motivation. You strive towards goals for personal satisfaction or accomplishment.

Some other important types of motivations are as follows

1. Achievement Motivation

It is the drive to pursue and attain goals. An individual with achievement motivation wishes to achieve objectives and advance up on the ladder of success. In this case accomplishment is important and not the rewards that accompany it. This type of motivation is more important for professionals.

2. Affiliation Motivation

It is a drive relate to people on a social basis. Persons with affiliation motivation perform work better when they are complimented for their favourable attitude and cooperation. This motivation is useful where money cannot motivate, such as in case of minimum wage employees and contingent professional.

3. Competence Motivation

It is drive to be good at something, allowing the individual to perform high quality work. Competence motivated people seek job masterly, take place in developing and using their problem-solving skills and strive to be creative when confronted with obstacles. They lean from their experiences. Specialists, like heart surgeons would feel motivated if they get chances to operate upon unique cases.

4. Power motivation

It is the drive to influence people and change situations. Power motivated people create an impact on their organisation and are willing to take risk to do so. Politicians usually function under this category of motivation.

5. Attitude Motivation

Attitude motivation is how people think and feel. It is their self-confidence, their belief in themselves, and their attitude to life. It is how they feel about the future and how they react to the past.

6. Incentive Motivation

It is where a person or a team reaps a reward from an activity. It is "you do this and you get that" attitude. It is the type of reward and prizes that drive people to work harder. Most of the unorganised job workers get motivated when they are offered more money.

7. Fear Motivation

Fear motivation threatens a person to act against his will or face the consequences. It is instantaneous and gets the job done quickly. It is helpful in the short run. In army this kind of motivation is very popular.

Importance of motivation

Motivation is one of the most important factors in determining the level of success that a person will achieve. Motivation is required throughout the entire goal achieving process. The following are some key reasons why motivation is so important.

1. Identifying a goal

Goals come from a desire to change something in your life. You may be motivated by positive or negative factors. Maybe you are unhappy with some aspects of your life. If this is the case, you would identify what you want instead and set a goal to achieve it. Alternatively, there may be something which you enjoy in life but you would like to experience more of it or, there may be something which you would like to add to your life.

Whether it is a desire for change, a desire to add more, or a desire to add something new, it is motivation for improved circumstances which inspires the goal.

2. Prioritization

Whatever you want to achieve in life, you will need to commit yourself to it and focus your energy on it. There is a limit to everything that you can do, as time and energy are limited resources. As you are constantly flooded with things that you could do, you need to be able to *priorities your activities and focus on the most important tasks*. This allows you to achieve best results. Motivation is critical in helping you to determine which activities should be prioritized.

3. Action

Anybody can draw up a plan, but the value of that plan will never be known until you put it into action. Setting a goal is not enough to achieve it. You need to take consistent, decisive and effective action to bring that goal to reality. The difference between those who consistently take action towards their goals and those who do not, is motivation. There will be days when you do not really want to do anything. On these days, it is your high levels of motivation that will push you through.

4. Overcoming setbacks

This can be applied to all goals. How many times have you given up on something just because you hit an obstacle? Almost any goal worth achieving is going to require some capacity to recover from the difficulties. You will hit obstacles and when you do, you must decide whether you will continue or give up. Motivation will ensure that you keep going.

Jack Canfield reported that 94 per cent of all sales people give up on the sale after the 4th time of asking, but 60 per cent of all sales are made after the 4th time of asking. That means that 94 per cent of sales people miss out on the chance of making the sale because they do not have the motivation to keep asking.

5. Staying the course

Most important goals will require a great deal of time to accomplish. It is easy to lose motivation and lose sight of your objectives'. If you allow this to happen, you may never reach your end goal. Highly motivated people keep on taking consistency action towards their goals and they also strengthen their motivation on a regular basis. A clear vision for life and, the effective goals to realize that vision, are essential if you want to live the best life. However, goals and a vision are only dreams unless you take consistent action to bring them to life. Motivation is a must to take this consistent action. No matter what goal you set, never lose sight of your need for motivation as it will be the one thing that will push you through the tough times.

Importance of Motivation and Educational Environment

Self motivation in the students

Motivation has several effects on the learning and behaviour of students. Motivation directs behaviour toward particular goals. Motivation will increase the time devoted on the task and is also an important factor affecting learning and achievement of students. Motivation enhances conscious thinking process.

1. Attach meaning to your studies

Even of someone is forcing you to go to school or you feel you don't have a choice, change your attitude and you will find the task much more pleasant. Take personal ownership of your knowledge and learning. Concentrate on learning, not just classes or studying for exams. Envision far in the future how your studies may benefit you, also others and God. Do not wait for someone to force you to study.

2. Create a plan: map your semester, your week, and your day

Make good use of monthly or semester calendars, daily planners, weekly time management worksheets, and electronic reminders that you might set up on your cell. Phone or computer. Put all due dates and other critical dates for the semester in a place where you will see them frequently. Take a few minutes in the evening to plan out what you intend do to next day. At the end of each day, take inventory on what you finished and what is unfinished. Celebrated your accomplishments.

3. Build a routine

Once you know your schedule of classes and other commitments each week, put these on a weekly time management spreadsheet that shows every hour of every day for the week. First add in your class times, sleep, exercise, etc and then find blocks of time where you can study. One week might look very similar to the next, and this is the way that you get into a routine. Having a routine, helps life to go smoother and easier. It allows you to be more organised and productive.

4. Identify several comfortable working/study environments

A good study environment should be quiet with very few distractions and interruptions. It should be comfortable and well-lit. You might have a main study base, but then you should also develop two or three alternates. For example, your room might be the main base but you might also find a favourite spot at the school library and then a third spot at the coffee shop. For some people, it is helpful to study in one spot for an hour or two, and then change to a new place. You can determine how frequently you need to change in order to keep your attention and motivation a high level.

5. Sleep, Eat, Exercise

Successful students have to stay healthy in order to perform effectively, just like an athlete. There is a huge connection between physical and mental wellness. Be sure to get the right amount of sleep (not too much or too little). Eat real food, not junk food. Enjoy some exercise every day (even just walking or biking is good). These three things keep your mind sharp and your spirits up.

6. Tame the time monsters (Internet, video game, porn, social drinking, etc.)

There are some things that suck the time right out of your weekly schedule and may leave you feeling tired and discouraged. Hours can pass quickly when you are lost in video games. These time monsters can also become regular bad habits. Exert some self-control to limit the impact of these distractions in your life. If you find yourself unable to cut back, seek help.

7. Avoid multi-tasking

We live in a culture that teaches us to do about things at the time. This common tendency actually creates a high amount of stress and leaves us feeling restless. Multi-tasking is a bad news for studying. Choose a subject or task and focus only on it. In order to accomplish this, you will probably need to shut off the cell phone, close up Face book, and put away your email.

8. Take planned and earned breaks

Even the most dedicated person needs a break from time to time. Breaks actually helps us become fresh and renewed for the next, push forward. Set a time limit for your study followed by a break. Your break might involve walking form one study location to the next. Breaks can also be a way of rewarding yourself for being focused. An exercise break from studying actually helps accomplish two good things at the same time. A break for prayer and meditation, is another great way of recharging mentally and spiritually.

9. Connect with a support system

Gravitate toward friends that study hard and want good grades. This way you can encourage one another and be accountable to them, working at the same goal. If you hang out with friends who just want waste time, then guess what you are likely to get?

10. Talk positively to yourself

Our attitude is critical, and the things that we think and believe, shape our feelings and motivations. It is better to tell yourself things that you truly believes, but you might have to fake it at first. If I tell myself "I am overwhelmed by this class and am sure to fail" then what do you think is going to be the result? To yourself, for example, that you are going to have a good, focused study period and then start the study. Recall the good grades that you have made in the past. Become convinced that you are smart enough to learn what you need to. Muster up the courage to do your best.

The expert in everything was once a beginner.

■■■■

25

Sensitivity

MEANING AND DEFINITION

Sensitivity is the strength of the capacity to detect and discriminate stimuli. It denotes the strength of the perception of a stimulus, in an individual. A person who has higher sensitivity will perceive a stimulus more strongly than someone who is less sensitive to the stimulus. If two people are looking at a light that is increasing in brightness, the one who is more sensitive will squint and close his eyes sooner than other person who is less sensitive. People have variations in sensitivities to different stimuli. For example, research has shown that redheads are more sensitive to pain than people with other hair colours.

Sensitivity has many shades of meaning, but mostly relate to your response to your environment; either physical or emotional. It is a term with wide ranging implications depending on the context it is used.

However, with respect to the human beings in general:

- Sensitivity is defined as an ability to understand what other people need, and to be helpful and kind to their feelings liable to be offended or hurt: or their sensitivities.
- Sensitivity is the quality or condition of being sensitive.

Sensitivity can be described with reference to various different factors or conditions such as:

Kindness – an ability to understand what other people need, and be helpful and kind to them.

Nature – the quality of being easily upset by the things people say or do, or causing people to be upset, embarrasses, or angry.

Reacting – having a strong physical or emotion reaction to something.

Social Sensitivity

Social sensitivity can be described as the proficiency at which an individual can identity, perceive, and understand cues and contests in social interactions along with being sociality respectful to others. This is an important social skill and having high levels of social sensitivity can make you more well-like and successful in social relationship.

An example of someone with low social sensitivity would be an individual, who only talks about themselves, interrupts and talks over others, and who ignores social hints to stop talking.

An example of someone with high social sensitivity would be a person who understands conversational cues and stops talking in order to listen at the appropriate time.

What Is Hypersensitivity?

Hypersensitivity also known as being a "highly sensitive person" (HSP) - is not a disorder. In fact, it brings many benefits, such as being able to "read" the mood of a room quickly and factoring in subtle cues when making a decision. "It is good in some situations and not in others," says psychologist and psychotherapist Elaine N. Aaron, Ph.D, author of The Highly Sensitive Person. She believes knowing that you have hypersensitivity is important. People with Attention Deficit Hyperactivity Disorder (ADHD), may have trouble focusing on some tasks and subjects.

How highly sensitive people can cope up with today's over stimulated world.

According to Zeff following are the tips for coping up with high sensitivity.

1. Set a bedtime and morning routine

For at least an hour or two before bedtime, shut down all electronic equipment and engage in calming activities, such as reading an interesting book. Keep calm in the morning too. Spend 30 minutes centring you by practicing yoga or meditation.

2. Identify your triggers

As all highly sensitive people are different, so it is important to determine what stimuli trigger your discomfort.

3. Plan ahead

If you are sensitive to loud noises and crowds, plan your activities so as to avoid noisy or crowded places and occasions.

4. Work around triggers

Planning ahead does not mean avoiding the activities you love.

5. Investigate current stressors and solutions

If you are in a super stressful job, consider what causes the stress and whether you could really do something about it. Discuss with others to find a solution and try whatever you can think of to improve the situation, before you decide to quit.

6. Remember your gifts

Even though being highly sensitive is not a flaw, you still might feel and that you are easily bothered by things that others are not. But being HSP also tends to have many positive qualities, including being creative, conscientious, loyal and deeply appreciative of the arts. Sensitively is the should of every creative activity or art.

7. Take mini retreats

Zeff has stressed the importance of downtime. He suggested, getting away at least once a month and relaxing several days a week. Enjoy nature (if you live in an urban area, visit a park) or get a massage. Add calm into your week with activities like aromatherapy.

Periodical relaxation helps in excelling in performance when you get back to regular business.

8. Engage in gentle exercise

Yoga, physical exercises and walking are always good for your physical and mental health. It is also better to exercise before 6 p.m. or 7 p.m. because it takes a few hours for your nervous system to calm down.

9. Speak up

Non-HSPs simply do not notice loud noises or strong. Smells or other stimuli that might be bothering you, so speak up. For instance, say your co-worker talks loudly on the phone. If you think they will be open to adjusting their behaviour, first build a rapport with them. Then explain that while they are not doing anything wrong, you have a trait that makes it tougher to tune out stimuli (which about 20 percent of people have). You do not want to interfere with their lifestyle, but may be they could speak more softly or when you are on break.

■■■■

26

Success

MEANING AND DEFINITION OF SUCCESS

In order to lead the life that you desire, you must set your own goals and idea of success according to what you want, and not what your parents want or anybody else advises or someone has already achieved a particular goal. This is all about you, your life, and your own idea of success.

The dictionary described success as "attaining wealth, prosperity and/or fame". Success is something that we all want to attain. It is a station of life that we dream about. Who does not want to be successful and powerful? You need to realize that chances of attaining success in life, in whatever you dream to accomplish, depends on your attitude and the principles you follow to make your dream a reality.

Had it been easy to achieve success, everyone would have attained it, but then it would not have been something special and rewarding. Opportunity always favours those who are best prepared. Luck happens when hard work meets opportunity, and success when you give it your absolute and best effort.

The purpose of defining your own success is to start living a life based on your own expectations and terms. The actions that bring you joy, an sense of accomplishment, and progress, are the areas where you would be investing your energy. Work towards daily success, not only the success that comes at the end of the journey. Do this every day, and your actions will bring you closer to your unlimited goals.

How to define success in life?

The only person that can answer the above question is you. Every person thinks differently about being prosperous in life and is defining success in some other way, so there cannot exist a single definition that is suitable for all. It is very important that you know exactly how to define success in life! Make yourself aware what accomplishment, success and prosperity in general means to you in your life. Some might define success as having luxurious cars and a huge mansions, whereas others consider a life full of joy and happiness with their family as the true meaning of success. Once you have figured out what is important for your personality, you are able to focus on your visions and goals.

The moment you define what the meaning of success is to you, the only way to achieve it, is to take the steps forward in accomplishing your goals or objectives. You need to stop trying to always please others and meet their expectations. But you must find joy in the things you are already doing or will start doing.

Definition of failure

The opposite of success is failure, as it means to fail while trying to achieve aims or objectives. Besides this regular definition of failure, it also can be said that even wealthy and successful persons fail in their lives. Just think about the rich and famous, and all their scandals, addictions and subdues. All of them were extraordinary persons, but a lot of them were also extremely unhappy with their lives and were not able to see the meaning of success. Wealth cannot be defined with money, but instead with values in your life that make you a happy person, such as friendship, relationships, and your family.

The definition of accomplishment

Accomplishment can be seen as the process to become successful and with every accomplished goal you take a step towards prosperity and a life full of success.

One of the most important key steps to achieving success in life is to know the meaning of success for your personal life. The true meaning of success goes far beyond the common definitions of success, such as having a lot of money, being wealthy, having a lot of material possessions and earned degrees. True success in life cannot be measured with these factors, but instead with the amount of people that are able to live a better and more advanced life because of what you created. This is the meaning of success. Media and society make us often conclude that living a successful life means to be extraordinarily wealthy and have a lot of belongings. But the meaning of success is to live a happy life and to make this world better places for everyone.

The difference between accomplishment and success

Accomplishment is often associate with success, but it is not the same. Accomplishment refers to the result we desire when we attempt to reach specific goals. Basically it is the result that we plan or expect to occur. Success is the positive consequences or outcome of an achieved accomplishment.

How to achieve success in life?

- The process of becoming successful starts with elaborated goal setting.
- Define a strategy and a plan how you intend to reach your goals, aims, and visions.
- Keep in mind that success is the consequence of having earned a series of accomplishments, so make sure to divide your goals into easier to reach subgoals.

Principles to attain success

These are the certain principles you need to live by, to attain the success you dream of:

1. Be grateful and appreciate what you have

The final success principle is to appreciate what you already have. Realize that by sheer virtue of the fact that you can be reasonably certain you will live from one day to the next, you already have enough. Appreciate what you have achieved so far. Being thankful and grateful will change your life dramatically.

2. Keep the right people around you

Negative influences and negative people will only hold you back. On your road to success, you need to keep all the negativity away. Keep people and ideas around you that will support your success and your belief in yourself. This is because you tend to feel and think life that people you move around with or you surround yourself with. Make sure you surround yourself with the most capable positive and supportive people who will help you to get where you want to reach.

3. Use affirmations

Affirmations and encouragement will keep you going along the long journey. It is the positives mindset of "I am going to make this happen". It fills you with a belief in yourself. With positive attitude you are defining your own reality. As your mind adopts this view, the positive energy generated helps in shaping the world around you, as you would like to fit in.

4. Set goals along the way

A final goal means setting the big picture and seeing your finished canvas. But your goals along the way are the paint that you use to get there. These subgoals should be specific, measurable, realistic, attainable and timely.

5. Clear vision

Success is about having a target. The problem is people do not know how to identify a target, so they are just shooting in the air, hoping to hit something. Having a target will allow you to aim where your need to hit. Even if your fall short of the target, you will still come closer to hitting the target.

Clarity is everything and understanding clearly what exactly you want, will bring success to you. Your efforts should not be directed towards a broad idea, but they should be focused and to the point. By knowing the destination you want to reach, you can constantly view your path and decide whether it is right course leading you to get where you want to reach.

6. Determination is everything

If you are determined you can do anything you set your mind to. It is about finding a way to make it happen. Where their is a will, there is some sort of a way. Strong will power can even create a way to achieve whatever you want. You need to worry when there is not will or a lack of determination.

7. Focus on growth

Obstacles and challenges will always be in your way during your journey. It is about figuring out how to overcome them and go ahead. Success is a problem-solving, solution-driven situation, and that is where growth comes from. Do not ever become too comfortable where you are.

8. Thing bigger than you are

If you look at the mindset of successful people, you fang it is all about thinking bigger. It is about having no limits to think big and believing that anything in this world is possible, if you put your mind to it. By thinking bigger, you are no longer limited by what you think is possible.

9. Use the powered of dreams and your imagination

What you see for yourself is what will manifest in your life. Dreams do come true, but this happens only when you work hard towards turning them into a reality. Your dreams can be fulfilled and your visions can be attained in your life, if you manage them, think positive and chase them.

10. Move in the right direction

Understanding the right directions and planning how to get there is important. Things are not going to move quickly, in fact they are going to be slow and take time, but as long as you are following the right direction, you will succeed. Success is a marathon long distance run, not a sprint or short race.

11. Massive action

People set goals, have dreams and want different things in life, but never get them because they do not actually start working toward them. They do not take the massive action that is needed. People love to push things to tomorrow, but tomorrow never comes. It is about starting today and putting in the action that is needed.

12. Do what you love

Money should never be your end goal or your motivation. It has to be something greater than that. It is your purpose and what you become in this world. Doing what you love to do will allow you to give your best effort and keep you excited while doing it. Find something that you enjoy doing, find your purpose and use your passion to propel you.

13. Preparation

Preparation is the foundation of success. The most prepared person has the best chances. That it the person who perfects his craft, has been throughout working hard and also loves what is being done. With preparation, your essentially create your own opportunities and that can take you anywhere you want to go. You need to have the right preparation to meet your goals because without it, you are taking an uncalculated risk.

Keep in mind the meaning of success and always try to remember how to define success in life so that you will be able to live a full life of happiness and joy.

Success Quotes

These are some very inspiring and motivation quotes that will accompany you on your journey to success and happiness:

- Action is the foundational key to all success – Pablo Picasso.

- Always bear in mind that your own resolution to succeed is more important than any other – Abraham Lincoln.
- Defeat is not the worst of failures. Not to have tried is the true failure – George Edward Woodberry.
- Discipline is the bridge between goals and accomplishment – Jim Rohn.

■■■■

27

Selfless Service

MEANING

Selfless service or Seva in Sanskrit is a service which is performed without any expectation of result or award for performing it. Such services can be performed to benefit other human beings or society.

Being selfless is always willing to give to others without for proposal gain. If you give time, money, or things to other people without expecting something in return, that is selfless attitude. Most people agree that the world would be a far better place if there were more selfless behaviour. Thought of reward or repayment. In ancient India seva was believed to help one's spiritual growth and at the same time contribute to the improvement of a community

So, is it really possible for an ordinary person to serve others without expecting reward or recognition, or does one need to be a saint to engage in such service? According to wellness expert Daniel Scott, the answer is that we all have the capacity for selfless service.

Through selfless service or the poor and needy and charity one can:

- Develop the heart and cleanse the mind.
- Purify the heart and make it a fit abode for God.
- Fill the heart with divide virtues.
- Attain the vision of God.

Research indicates numerous benefits to those who engage in selfless service, including reduction of emotional disturbance, greater longevity, stress reduction, improved morale, increased self-confidence and self-esteem, better healthy, pain reduction, and greater overall happiness.

According to Sri Ravi Shankar, the joy of giving is far more rewarding than the joy of getting, and can be compared to a thrill or high. Services are sacrifice and sacrifice is an act of giving out, which will cost you something.

Types, kinds of selfless service

- Physical – Services through sharing the physical strength of the body, sharing body or physical abilities for the benefit of the needy.
- Financial – Sharing acquired wealth and resources for humanitarian activities. Donates and charity for human welfare, healthcare, community development, disaster management etc.
- Intellectual – Expression of inner maturity creating awareness, Service intellectual needs, sharing knowledge, education to public, creating awareness of internal peace.

"Engage yourselves in service activity social service as service to God. This is the best way to earn God's Love. Love all and serve all. Your entire life will be sanctified thereby". Satya Sai Baba

Satya Sai Baba suggest a pure heart, charity, love, selflessness, and a passionate attitude as prerequisites for set: Phyllis Krystal, a devotee of Sathya Sai Baba writes in `Taming Our Monkey Mind', Service whether to family members or others, should be undertaken for the right reasons and not with any ulterior motives in mind, for that would curtain its effectiveness. To be free from ego it should flow out of deep compassion for the pain, hunger, feelings of rejections, loss, or any other problem from which either an individual or a group is suffering. Only then it will be benefited the one who serves as well as those receiving the services.

Therefore, one should not take up any particular service with an intention to gain recognition, feel important or worthwhile or any other reward from the beneficiaries', or for any kind or personal gain. Let's take a look at some great leaders starting with Nelson Mandela. He was a selfless leader who lived his life for his people and environment, yet he became one of the greatest who ever lived. During his life and after his death he was celebrated as a hero. Mandela has become a household name all over the world.

Mahatma Gandhi was another selfless leader recognized throughout the world, who cared the pains and interest of his people above himself and he also became great during his life time. Mother Teresa was another woman who rendered a selfless service and the Lord blessed her in all her endeavours.

Wangari Maathain, a great female environmentalist, political activist and conservationist from Kenya is one more example of a selfless leader who had so much love for her people and environment. She decimated her life time to political activism and environmental issues in her country and was greatly honoured.

Baba Amte was an India social worker and social activist. He was from a rich family and a renowned lawyer and was successfully practicing law. But he later chose to undertake selfless service of the society. He is known particularly for his work for rehabilitation and empowerment of people suffering from leprosy. He started an organisation for the leprosy patients known as Anandwan.

The best way to find yourself it to lose yourself in the service of others: Mahatma Gandhi.

■■■■

28

Case Studies or Ethical Life

Ethical living is the philosophy of making decisions for daily life, which take into account ethics and moral values. This includes living according to moral principles that govern the person's behaviour or conduct or any activity. The idea is to dedicate oneself for the observance of moral values and thus bring about a positive change in the life of self as well as others, which would ultimately result in the welfare, progress and upliftment of the society.

Albert Schweitzer describes certain key issues of Ethical Living which can be summarized as follows:

1. A man will not find peace until he includes all living things in his circle of compassion. One must accept and practice the philosophy of interrelation among all living beings.
2. A man is ethical only when life is sacred to him, this means life of all for which he should be ready to help. Helpfulness is prime code of ethical life.
3. Think about suffering of others, try to feel their pain and make attempt to give relief.
4. The first step in the evolution of ethics is a sense of solidarity and compassion for other human beings.
5. A truly ethical man always feels the compulsion to help others and avoids injuring or harming anybody. One of the most dignified qualities of the human being is expressed though the inherent will to help others.

6. Realize that every act committed by you has a valued wide effect in all spheres of life, you directly and indirectly participate in. The quality of your intention can improve quality of life of all beings on the planet.
7. Example is the only thing that can influence others. Lead by setting example and remain as a role model and guiding light for all generations.
8. Ethics is the activity that secures the inner perfection of personality. Recognize that your own perfection and spiritual evolution is connected with well-being of others. As we perfect ourselves as ethical humans, we shift the standers of spiritual and social success.
9. Know that the gentleness of your soul can be felt, and it expressed through acts of kindness, and selfless service.
10. High ethical and moral standard of behaviour includes understanding of humanitarian values and never sacrificing human being or human interests for consumerism, self interest or any sort of power.
11. No body should be treated as stranger and dialogue with people whom we do not know, is the beginning commitment to peaceful and understanding co-existence.
12. Always seek and maintain the dignity of individual human beings, their communities and the entire environment.

There are several examples of enlightened individuals who by their ethical living and noble values have set remarkable examples for others. The following are a few examples among the many that are worth following:

1. Mohandas Karamchand Gandhi (2 October 1869–30 January 1948) was the leader of the India independence movement against British rule. Mahatma Gandhi is universally accepted as an exemplary model of ethical and moral life, with a rare bleeding of personal and public life, the principles and practices, the immediate and the eternal. He believed in a single standard of conduct based on truth and Non-violence. He successfully led non-violent struggles against colonial rule, economic and social exploitations, racial discrimination, and moral degradation.

- Even one person can make a difference:
- Strength comes not from physical capacity but from an indomitable will;
- Given a just cause, victory is certain through nonviolence, capacity for self-suffering, and fearless;
- Leadership by example is the one most effective.

A central quality of Gandhi's leadership was its natural evolution through intense interaction with the people and the events. He was astutely conscious of his own imperfections. Gandhi single-handedly made non-violence a universal substitute for violence and the bed-rock of his leadership. His non-violence was the way to counter injustice and exploitation. Gandhi's leadership was a running ethical lesson to his followers as well as his opponents on `how to live' An outline of the basic ethical tenets of Gandhian leadership, are given below.

1. Truth
2. Nonviolence
3. Right Means and Right Ends
4. Primacy of Duties over Right
5. The Deed, not the Doer
6. True Religion (Universality and Brotherhood)
7. Aparigraha Or Non-possession (Voluntary Poverty)
8. Yajna (sacrifice and Service)
9. Satyagraha or Nonviolent Conflict Resolution

2. Mother Teresa

"Humility is the mother of all virtues; purity, charity and obedience. It is in being humble that our love becomes real, devoted and ardent. If you are humble nothing will tough for you, neither praise nor disgrace, because you know what you are. If you are blamed you will not be discouraged. If they call you a saint you will not put yourself on a pedestal." Mother Teresa

Agnes Goxa Bojaxiu better known as Mother Teresa, came from Yugoslavia to Calcutta in January 1929 and never left this country. Her aim in life was to serve the sick and the poor and she dedicated her full life toward this purpose.

She managed to open several institutions to help the downtrodden *e.g.* 'Missionaraies of Charity' Nirmal Hriday', and Shishu Bhavan'. Mother Teresa was awarded the Noble Peace Prize in 1979 and the Bharat Ratna in 1980.

Despite all the public acclaim national and international honours, mother Teresa remained humble, kind and generous till the end.

Mother Teresa gave these rules to her Sisters to help them develop humility:

1. Speak as little as possible about yourself
2. Keep busy with your own affairs and not those of others.
3. Avoid curiosity.
4. Do not interfere in the affairs of others.
5. Accept small irritations with good humor.
6. Do not dwell on the faults of others.
7. Accept censures even if unmerited.
8. Give in to the will of others.
9. Accept insults and injuries.
10. Accept contempt, being forgotten and disregarded.
11. Accept injuries and insults.
12. Be courteous and delicate even when provoked by someone.
13. Do not seek to be admired and loved.
14. Do not protect yourself behind your own dignity.
15. Give in discussions, even when you are right.
16. Always choose the more difficult task.

3. Dr. A.P.J. Abdul Kalam popularly known as the Missile Man of India and People's President is one of best examples of ethical living. Coming from a very humble background, he rose to become a renowned space scientist and president of India through his sincere, dedicated, moral and ethical life style. He has received honorary doctorates from 30 Universities and institutions and many awards including India's highest award Bharat Ratne. Simplicity was one of his greatest virtues and while occupying the highest position in the country. Abdul Kalam was not swayed by power, but used love and peace as tools to be one among the masses. He gave maximum importance to ethics and ethical values in the life of an individual as well as society for the progress of the nation. Dr. APJ Kalam defined human virtues and ethics in most functional manner.

- Individual characters ethics – Gratitude, courage, self-discipline, reliability, kindness, caring, and sincerity.
- Work characters ethics – Honesty, creativity, trustworthiness, coworker appreciation, task competition, empathy, cooperativeness, supportiveness, and tolerance.
- Professional character ethics – Self-regulation, trust, independence of judgment, and impartiality
- Total commitment is not just hard work, it is total involvement
- To succeed in your mission, you must have single-minded devotion to your goal.
- To put the heart and soul into the mission
- Willingness to use inner resources to invest in life, especially imaginations, will bring success.

He commented that:

- The nation needs to have ethics in all its tasks, for great economic prosperity and peace.
- If the nation is to have ethics; society has to promote ethics and value system.
- If society is to have ethics and values system, families should stick to ethics and value system.
- If families have to get evolved with ethics and value system, then the parent should have inbuilt ethics.
- Parental ethics come from great learning, value based education, and a creation of the clean environment that leads to righteousness in the heart.

4. Lal Bahadur Shastri was a man best known for his integrity, honesty, hard-work simplicity and austerity in Indian politics. In the history of Indian politics which is dominated by dynasty, corruption, wealth and raw power. Lal Bahadur Shastri stood out as an odd man by virtue of his ethics, integrity, simplicity and austerity. He was deeply influenced by Mahatma Gandhi and took part in the freedom struggle.

While he was jailed under the British, he was informed that his daughter was sick. He requested for 15 days to visit her and was granted under the condition that he must not participate in freedom struggle in those 15 days. Unfortunately by the time he reached his home, his daughter had died. After performing the funeral and last rites for 3 days. He immediately returned to jail inspire of having 12 more days of freedom from jail.

Similarly, in another event, while in jail, he was informed of his son's serious sickness and this time he got a permission of 7 days of leave from jail. He stayed with his son all the 7 days and although his son had still not recovered, Shastri went back to jail to keep up with his words! He served as the Railway Minister under Jawahar Lal Nehru. Such was his moral standards at his job that immediately after a train accident, he took moral responsibility and submitted his letter of resignation.

5. Kisan Baburao Hazare popularly known as **Ann Hazare**, is an Indian social activist who led movements to promote rural development, increase government transparency, and investigate and punish corruption in public life. Hazare served the Indian Army, where he initially worked as an army truck driver and later as a soldier. During the Indo-Pak War of 1965. Hazare was posted at the border in the Khem Karan sector and was the sole survivor of an enemy attack.

Inspired by the book by Swami Vivekanand "*Call to the youth for nation building*" besides other words of Vivekanand and Vinoba Bhave, he decided to dedicated his new life to serving people. He started with the development, a fragile ecosystem, neglect and hopelessness. He began his campaign with setting up a grain bank, watershed development, drive against untouchability and alchohalism. He worked to increase literacy rates and education levels. He campaigned between 1998 and 2006 for amending the

Gram Sabha Act, so that eventually have a say in the village's development. The state government initially refused, but eventually gave in to public pressure. It became mandatory to seek the sanction of the Gram Sabha (an assembly of all village adults, and not just the few elected representatives in the gram panchayat) for expenditures on development works in the village.

In 1991 Hazare launched the Bharashtachar Virodhi Jan Andolan (BVJA, People's Movement against Corruption), a popular movement to fight against corruption, in Ralegaon Siddhi. This fight against corruption has become his life goal, for which he suffered hardships, was imprisoned, but his determination and tireless sustained efforts have established him as one of the most recognized leader against corruption at the national level. His major efforts include Right to information movement. Lok Pal Bill movement, Electoral Reforms movements. A true follower of Gandhian philosophy, he professes non-violence and uses hunger strike as his main instrument to achieve his mission

6. Anand Kumar is an Indian mathematician and a columnist for various national and international mathematical journals. He is best known for **his Super 30 programme**, which he started in Patna, Bihar in 2002 for coaching economically backward students for the entrance examination for the Indian Institute of Technology (IIT's). By 2017, 396 out of the 450 has made it to IIT's. He has been invited by the prestigious Massachusetts Institute of Technology (MIT) and Harvard University to speak on his globally acclaimed effort to mentor students from the underprivileged sections for admission to IIT.

In 1992, Kumar began teaching Mathematics. He rented a classroom for Rs. 500 a month, and began his own institute, the Ramanujan School of Mathematics (RSM). Within the span of year, his class grew form two students to thirty-six, and after three years there were almost 500 students enrolled Ramanujan School of Mathematics holds a competitive test to select 30 students for the `Super 30' program. Many students appear for the competitive test, and eventually thirty intelligent students from economically backward sections were selected. Anand Kumar, tutors them, and provides study materials and lodging for a year. Whole programme is conducted as a dedicated social service without any commercial interest. His mother, Jayanti Devi, cooks for the students, and his brother Pranav Kumar takes care of the management.

Anand kumar has no financial support for Super 30, from any government or private agency, and manages on the tuition fee he earn from the Ramanujam Institution. After the success of Super 30 and its growing popularity, he got many offers from the privates, both national and international companies, as well as the government for financial help, but he always refused. He wanted to sustain Super 30 through his own efforts.

■■■■

29

Positive Spirit

Positive spirit can be best defined as the optimistic attitude that comes from knowing, that God works on all things that are good for those who love him. Being optimistic is to always look on the brighter side of things.

Negative attitude not only prevents you from fully enjoying your life, but can also have a significant impact on your environment. One of the best things you can offer to your family, organisation or community, is your positive attitude

In the wise worlds of Michael Jackson, "If you want to make the world a better place, take a look at yourself and make a change". By making small adjustments is your daily life and mindset, positive habits can be formed. These will make yourself happier and also everyone around you.

A **positive attitude** does not mean ignoring the troubles of life. It just means being an optimist-looking for the good in everything, rather than being a pessimist-concretizing on the bad in every things. A Positive attitude leads to Success and Happiness. *Positive attitude* helps you cope more easily with the daily affairs of life. It brings optimism into your life, and makes it easier to avoid worries and negative thinking. Sometimes your attitude can make all the differences in the world.

Here is a list of 10 habits that can run a gloomy outlook into an effortlessly optimistic attitude.

1. Keep gratitude in mind

Sometimes one single bad event can ruin an entire day and an unpleasant happening at night can overshadow the next day. We are aware that our mind tends to hang on to the negative and we get worried easily. To avoid this we can intentionally focus on the good parts of our day. Try writing down five things that you feel grateful for every day, and see how your attitude changes. Science has found that attitude can significantly increase your happiness, and it also protects you from stress, negativity, anxiety and depression.

2. Reframe your challenges

No one can win or succeed every time. "Sometimes you win and sometimes you learn." – Robart Kiyosaki. There are no dead ends, only re-directions. Despite our best efforts, there are only very few things in life that we can completely control. Therefore, we should not worry about things over which we do not have any control. What we can control is the effort that we put in and when we give our full effort, there is no reason for regret. Enjoy the challenges, do not resist or try to escape. Accept them as opportunities or adventures, instead of attempting to resist or avoid.

3. Get good at being rejected

Rejection is normal and is a part of life. No one can get along the walk of life without being rejected ever. Do not let it hearten you or dishearten you. Keep your efforts going. Do not expect the worse. If you wait for bad things to happen they seem to follow, and chances are you may miss good things along the way.

4. Use positive words to describe your life

The words that we use have a lot more power than what we think. How you talk about your life is how your life will be. This is so because your mind hears what you say. If you describe your life as boring, busy, chaotic, that is how you will perceive it and feel the negative effects in your body and mind. If you use the words simple, involved, familiar or lively, you will see your life in a different lift and find more enjoyment in the way you chose to shape your life. A study from US data suggests that having a positive attitude had a direct effect on your happiness.

5. Replace 'has' with 'get'

Have you ever noticed how many times we say that we have to do something? I have to go to work. I have to go for grocery shopping. I have to pay my rent. Now change this one little word to get and see what happens. I get to go to work. I get to go grocery shopping. Even, I get to pay my rent. Your attitude quickly changes from doing a duty or fulfilling an obligation, to get happily used to do your regular jobs as routine and enjoy doing everything you do. Try to make this change and you may feel and appear happier, and less stressed.

6. Do not let yourself get affected by people's complaints

Do not pay attention to complaints made by anyone, because this may lead to your thinking in the same way. A study done at the Warsaw School of Social Psychology shows that complaining leads to lowering of mood and negative emotions, decreased life satisfaction and optimism; emotional and motivational deficits.

7. Breathe

Our breath is directly connected to our emotions. Have you noticed we hold our breath sometimes when we are concentrating on something? Can you feel your breath change. When you are angry or anxious? Our breath changes, depending on how we feel. The great news is what the connection goes the other way too. We can also change how we feel, using our breath!

8. Be hopeful and expect good happenings even in tragedy

It's hard to have hope and to be positive in difficult times. However even in tragedies like violent disturbances we find people rising up and do their best to prevent them. During every natural disaster, major accident or any traumatic event, there are people who come forward to help the needy and they reach out to the sufferers with love and compassion. Hold onto the stories of modern day heroes and selflessness, in the times of fear and devastation.

9. Have solutions when pointing out problems

Being positive does not mean that you have to be indifferent to problems or just avoid them. Positive people have constructive criticisms to improve conditions. If you are going to point out problems in people or situations, make as much effort into suggesting solutions. Instead of pointing out all of the things that are wrong, offer ways to make them better.

10. Make someone else smile

Who do you think about most of the time? Honestly, most of us would say themselves. It's good to hold ourselves accountable, take responsibility for our life roles, hygienic, food, etc. But set a goal for each day to make someone else smile. Think about other's happiness and it will help us to realize our immense impact that our attitude and expression has on the people around us.

Comparison of Positive and Negative Attitudes

Sl.No	Positive	Negative
1	Always active ready to go	Passive and nervous
2	Strong will power and motivation	Tired and sluggish
3	Fully Concentrated and focused	Divide and indecisive
4	Feeling secure and in harmony with self	Insecure, out of balance
5	Light in movements	Heavy and slow
6	Always in happy and good mood	Very low and bad mood
7	Free from conflicts	Full of problem and conflicts
8	Strong with great self confidence	Unsure, lacking self confidence
9	Free from performance anxiety	Afraid, not daring to go for it
10	Bold enough to go well and aiming high	Anxious about performance

What is the difference between an obstacle and an opportunity? Our attitude towards it. Every opportunity has a difficulty, and every difficulty has an opportunity. "Cultivate an optimistic mind, use your imagination, always consider alternatives, and dare to believe that you can make possible what others think is impossible.

A possible attitude in the workplace can help you, whether you are a business owner, an employee, or a manager in any business environment. You will enjoy doing your work more and achieve your goals more easily and faster, with a positive attitude.

Cynicism (being suspicious) and sarcasm (hurting with words), are very common at every workplace, and these make you and your co-workers fail to achieve the desired goals. You may not be able to control the mood of the entire workplace; but you are always in control of your own attitude each day. Use these ten tips for developing a positive attitude in the workplace.

1. **Gratitude** – Appreciate everyone who does a good job and does it on time. Specially praise them who achieve more than they were required to do. Appreciation always has good effect on both the giver and the receiver.
2. **Celebrate the Victories** – For some reason, many people tend to focus on everything that is going wrong. Instead of worrying about or complaining wrong things, focus your attention on things that are going well. Celebrate the victories at work. Reward yourself or praise others for a job well done. If you are part of a team that does something successfully, arrange a get-together to celebrate it.
3. **Turn Problems into Opportunities** – If problems are encouraged, do not just focus on them. Instead, change your focus by asking yourself or your team, "How can we solve this issue, and what can we learn from it?" Everything, problem or challenge, is an opportunity to learn and grow.
4. **Keep Healthy and Well Rested** – It is tough to have a positive attitude when you are feeling unwell and are tired. Make sure you get enough sleep, eat lot of fresh fruits and vegetables, drink enough water, and get regular exercise. This will help in having much more energy and focus, and you will actually be able to do more in less time. You will also feel much better mentally, which will help you to maintain that positive attitude.
5. **Smile** – Force yourself to smile, even if you do not feel like. A smile will actually shift how you feel internally. When others see you smiling, they too feel better.
6. **The "No Complaining" Policy** – No one like a complainer. If you are a manager, tell your team that you would not like them coming to you with an issue, unless they have a proposed solution for the same. Go to your boss and explain that you have a challenge and would like to offer a

way to solve it. This minimizes complaining an turns the focus to possible solutions instead, thereby maintaining a constructive work environment.

7. **The Win-Win Policy** – Make it your own personal philosophy or make it a requirement in your workplace that each agreement reached must be win-win. Agreements and negotiations must never favour one party. There in always a way to make agreements positive for both parties involved.
8. **Put People in Position that use their Strengths** – People will be more motivated and give better results when they enjoy what they do. Different people have different qualities. Observe the strength of individuals and put their skills to use. If you utilize strengths of people, you'll see a dramatic shift toward a positive attitude in the workplace.
9. **Trust and Evaluate People Based on Their Results** – If you are a manager, trust your people to do the jobs. Give them authority to decide how to do their jobs, you will get amazing results. Tell you people what you want them to achieve, and then let them go to do it. When you trust them to do it well, they will honour that trust. Most people will do the best they can to live up to the confidence you placed in them, and they will appreciate you for it.
10. **Focus on What You Enjoy. Improve the things You Don't Enjoy** – Make a list of the things you enjoy about your job and your workplace. Make it a point to review things list often as a reminder that there are good things about your workplace. For the things you do not enjoy, ask yourself how you could make them tolerate. Could you do them at a different time, in a different way? Can you speak to your manager about them, and offer a possible solution? Can you see anything good about them? If so, focus on that instead do whatever you can think of, to make these tedious tasks more enjoyable.

"Train your mind to see something good in everything" – Buddha.

■■■■

30

Body, Mind and Soul

It is now becoming common in to use the terms **"body-mind-soul"** or **"body-mind-spirit"**, while referring to out totally as a human being, in various spheres like psychology, religion, holistic medicine etc. We have three aspects of our humanness, a body, a mind and an immortal identity-soul or spirit. Mind, Body, Spirit means that our well-being is defined not only by our physical health, but also form mental health and spiritual health. The mind forms a link between the soul and the body, and the two interact on each other. Today, we accept that there is a powerful mind-body connection through which emotional, mental, social, spiritual, and behavioural factors can directly affect our health.

Simple Tips to Balance Your Mind, Body and Soul

When we think about health, typically the first things that come to mind are diet and exercise. However, good health is not just about the physical body. Our mind and body are interconnected and affect each other tremendously. For example, a stressful situation causing negative thoughts can lead to physical pain or illness. It is important to maintain a healthy balance between your mind, body, and soul by nurturing your whole self, including your physical, mental, emotional, and spiritual needs. There are many things you can do in your daily life to achieve overall wellness. The following are simple ways to begin developing a mind-body-soul balance.

1. **Exercise:** Find an exercise you enjoy and start doing it regularly. Exercise is important for heart health, physical

stamina, and mood. Get at least 15 minutes of moderate to fast-paced exercise each day. Walk or ride your bike to cover distances. Yoga is amazing for your overall health. It helps you build strength, coordination, and flexibility while calming your mind. It also develops the mind. It also develops the mind-body-soul connections.

2. **Mediate regularly:** Mediation improves memory, attention, mood, immune system function, sleep, and creativity. All it takes is a few minutes a day to start reaping the benefits. Set sometime a side each day to rest your mind.
3. **Be grateful:** Take some time each day to think about the things you're grateful for, like family, friends, pets, food, shelter, health, or the beauty of nature. Think about the people, things and situations that have been good, appreciated them and be grateful. Gratitude brings peace and harmony to the mind, makes us relaxed and free from worry.
4. **Get plenty of sleep:** Sleep is regenerative for your body. The more sound sleep you get the better you will perform the next day.
5. **Avoid sitting for extended period of time:** Try to stand or move around while you work, if possible. Too much sitting is linked to heart disease, diabetes, and a shortened lifespan.
6. **Breathe deeply:** Whenever you can just stop and take a deep breath. Over time this will become a healthy habit.
7. **Ground yourself:** Keep you bare feet on the earth, as often as possible.
8. **Eat organic**: Chemicals are killing pests on the crops. They are not good for you either.
9. **Smile more**: It feels great.
10. **Spend more time with loved ones**: In our busy lives we need to make time for the people who matter to us most.
11. **Find a career path that is meaningful to you:** Chase your dreams, not riches. Read and learn often. Your education shouldn't stop once you're out of school. Open your mind

to new possibilities, beliefs, and interests by reading, talking online classes, watching documentaries, and attending workshops.

12. **Live your passion**: Do more of what you love. Set aside time each day to do what makes your soul happy. Many of us work so much that we forget how great it feels to paint, dance, make music, write, garden, or swim.
13. **Get outdoors more:** Go for a hiking, boating, picnics, outdoor sports and enjoy nature much more.
14. **Eat plenty of greens:** Dark leafy greens are rich in vitamins, minerals and chlorophyll. They help to alkalize the body. Add more plant-based foods to your diet. Eating last of vegetables and fruits can help to prevent chronic disease. Slowly start incorporating these ideas into your daily routine and see how they help in balance of body-mind-soul and have positive impact on your overall health.

■■■■

31

Attachment/Bonding and Detachment

Attachment with respect to human beings can be best described as affection, fondness, or sympathy for someone or something. Attachment is an attraction or a like which we express in many different ways to people, situation an possession.

In fact, we do not know how to live without attachment. The emotional bond that typically forms between infant and care giver is the means by which the helpless infant gets primary needs. It then becomes the means of subsequent social and emotional development. The early experience of the infant stimulates growth of neural pathways that will decide his patterns of response to many things.

The attachment experience affects personality development, particularly a sense of security. Research shows that it influences the ability to form stable relationship throughout life. Neuroscientist believe that attachment is such a primal need that there are networks of neuros in the brain dedicated to setting it in motion and a hormone to foster the process, oxytocin.

Role of attachment in life

- Attachment system provides first coping system of infant
- It sets up in the infant's mind a mental representation of the caregiver.
- It is wholly portable and can be summoned up as a comforting mental presence in difficult moments.

- It allows an infant to separate from the caregiver without distress and begin to explore the world around.
- Attachment contains within it the platform, the child's ability to survive independently.
- Attachment is the source of all our pleasures now. We are attached to our friends, to our relatives; we are attached to our intellectual and spiritual works. We are attached to external objects, so that we get pleasures from them.

However, what bring misery is but this very attachment. Whenever we are forced to be away from any person, place or situation we are attached to, we become miserable. We need some time and efforts to come out of such condition and we have to detach ourselves to be normal and joyful.

Attachment becomes unsafe when it entwines our emotional, physical and mental energy with another person

We become obsessed with that one relationship. Sometimes we expect another person to meet our every need or may be we expect that of ourselves for others. Trying to meet every need of another person is an impossible super human task that sets us up for failure.

When attachment becomes unhealthy, we may feel the need to detach from a relationship

Detaching can begin with setting a boundary in a relationship. It may means saying no to someone you have always said yes in the past. If you feel the need to agree with everything and everyone, it will naturally be uncomfortable when you voice an opinion that differ from your friend/spouse/child/parent.

Healthy attachment is the act of both holding and letting go

Healthy attachment is holding the bird with cupped hands, so it is free to try to fly, but safely being held. The freedom to fly is important, but the safety of being held is a part of the process, too. In detaching from unhealthy relationship, we often go to either extreme of over protecting or releasing too quickly. We must find the balance.

Detachment

Detachment also expressed as non-attachment, is a state in which a person overcomes his or her attachment to desire for thing, people or concepts of the world. Detachment is a very strong word. Our understanding of detachment is to repulse something to which we are attached or attracted.

Importance of the term

Detachment as release from desire and consequently from suffering is an important principle, or even ideal, in the Bahai faith, Buddhism, Hinduism, Jainism, Stoicism and Taoism.

The Bhagavad Gita, the basic text on the practice of detachment, is wonderfully explicit on this point. Krishna tells Arjuna that acting with detachment means doing the right thing for its own sake, because it needs to be done, without worrying about success or failure.

You might be surprised to hear that detachment is an important quality for relationships. But in fact, it is much more than that. Detachment is a highly beneficial state of mind in all the fields of life and in our relationship with objects, even our own physical body.

Detachment is not indifference, you might think that a detached person is one that has no human relationship al all, or if he does, then they must be cold and superficial. Detachment has nothing to do with in difference. In difference means a lack of interest and sympathy towards a person or object. Detachment, on the other hand, refers to the state of mind of being objectives and not clinging, and it springs from a deep consideration of the conditions of human existence.

Detachment in real life does not mean in difference: on the contrary, it will empower you to live every relationship with love and intensity, knowing that it could end at any moment.

Detachment is a state of mind that will help you both in times of joy and sorrow. We all know that life a mixture of pleasure and pain, of comfort and hardship. Probably you tend to cling to pleasure, hoping that it will never leave you, and your are over whelmed by pain, fearing that it will never end. But when you start practicing detachment, you will be able to ensure difficult moments with patience. In the same way, you will enjoy the beautiful moments of life without being tainted by the fear that they will end – as they undoubtedly will.

Being detached from success and failure, from pleasure and pain, will bring you back into connection with the only things that is invariably present, stable, and safe your centre of pure awareness and pure love. Detachment is necessary to reach a state of equanimity, and to be able to live fully in the present moments, without worries or regrets.

Learning to practice detachment is one of the most important tools to develop unconditional love, a non-clinging attitude towards both things and people, and the capacity to enjoy the present moment with intensity.

Hinduism

The Hindu view of detachment comes from the understanding of the nature of existence and the true ultimate state sought is that of being in the moment. In other words, while one is responsible and active, one does not worry, about the past and future. The detachment is towards the result of one's actions rather than towards ever things in life. This concept is cited extensively within Puranic and Vedic literature, for example:

One who performs his duty without detachment, surrendering the result unto the supreme lord, is unaffected by sinful action, as the lotus is untouched by water – Bhagavad Gita 5.10. Vairagya is a Hindu term which is often translated as detachment.

Jainism

Detachment is one of the supreme ideals of Jainism, together with non-violence. Non-possession/non-attachment is one of the Mahavratas, the five great vows Jain monks observe. Suffer heavy loss. A contented man is rarely disappointed he who knows when to stop does not find himself in trouble. He will stay forever safe.

How to Detach: 5 steps

1. Observe your mind

What kind of thoughts you usually think. What things or factors do you identify with most? Become a student of self and heighten your awareness of where attachment happens more frequently for you.

Recognize attachment comes with an emotional charge. Notice where you feel this in your physical body. It's different for each individual and learning our patterns is a useful tool in creating change.

2. Distinguish between the voice of your ego and the actual situation

Your ego might tell you that not getting the job you won't have ruined your career. The actual situation is: you are disappointed because you don't have something you never had in the first place. There has been no loss. Nothing has changed except your thoughts about your future potential. The actual situation is the same as it was prior to not getting the job.

3. Embrace uncertainty

Only a willingness to embrace the unknown provides security. As Deepak Chopra says, those who seek security in the exterior world chase it for a life time. By letting go of your attachment to the possibilities. This is where you will find true happiness, abundance and fulfillment.

4. Meditate on it

Meditation is a vehicle to help your mind release pattern of thought and action that no longer serve you. Spend some time in meditation each day and watch how the patterns in your life begin to change.

5. Don't beat yourself up for falling in to old habits

The first step in making change is recognizing what it is you want to change. Instead of getting frustrated or disappointed when you repeat the pattern of thought or habit. In time, this will allow you to transform your behaviour.

32

Spirituality Quotient

The term 'Spirituality' comes Latin, word 'Spiritus' which means "Spirituality" originally developed within Christianity, refers to a life oriented toward the Holy Spirit. Traditionally, spirituality refers to a religious process of reformation which "aims" to recover the original shape of man. Each culture defines spirituality in its own way abiding by a set of religious principles, yet it has nothing to do with religion. Growth of secularism has given rise to a broader view of spirituality, with wide ranging implications in personal, practical and professional life of human beings.

Modern spirituality typically includes a belief in a supernatural realm or supreme controlling power, beyond the known and observable forms. Spirituality denotes the mental aspect of life, as opposed to the material and sensual aspects.

Spirituality is a dynamic and intrinsic aspect of humanity through which persons seek ultimate meaning, purpose, and the transcendence, and experience relationship to self, family, others, community, society, nature and the significant or sacred. Spirituality is expressed through beliefs, values, traditions, and practices.

Spiritual practices

Mainly four forms of spiritual practices have been described by expert:

1. **Somatic practices:** Especially deprivation and diminishment. Deprivation aims to purify the body. Diminishment

concerns the repulsion of ego-oriented impulses. Examples include fasting and poverty.

2. **Psychological practices:** Like meditation
3. **Social practices:** Such as practice of obedience and communal ownership, reforming ego-orientedness.
4. **Spiritual practices:** All practices aim at purifying ego-centredness, and direct the abilities at the divine reality.

Spiritual practices may include mediation, mindfulness, prayer, contemplation of sacred texts, ethical development. Love and/or compassion are often described as the mainstay of spiritual development, Spirituality lays emphasis on the value of thoughtfulness, tolerance for practices and beliefs, and appreciation for the insights of religious communities, as well as other sources of authority.

Spiritual experiences

Neuroscientists have examined brain functioning during reported spiritual experiences finding that certain neurotransmitters and specific areas of the brain are involved. Experiments have also successfully induced spiritual experiences in individuals by administering psychoactive agents while spirituality can also be dampened by electromagnetic stimulation of the brain.

Among health-care professionals there is growing interest in "spiritual care", to complement the medical – technical approaches and to improve the outcomes of medical treatments.

Spiritual Quotient

Spiritual Quotient (SQ) is describing as a measure that looks at a person's spiritual intelligence in the same way as intelligence quotient (IQ) looks at cognitive intelligence.

Spiritual Quotient is study of Social, Investigative, Artistic, Realist, Contractor and Conventional aspect of the personality to assess how balanced a person is. Now-a-days, many authors, scholars claim that spiritual quotient is the ultimate intelligence of a human being and this intelligence decides the ability of an individual to solve problems on values and to take value based decisions.

Spiritual intelligence is useful to

- Transform self,
- Heal relationships,
- Cope up with adverse situations,
- Move beyond already conditioned habits of persons

To develop high **SQ**, each person's needs to approach the task according to his/her personality. Ego is one of the main obstacle standing one's spiritual progress and advancement. Great spiritual scholars from India suggested limiting and surrendering our ego, as when we surrender our ego, there is tranquility and peace in our interaction/dealing with the world.

Spirituality quotient (SQ) and its effects on personal and professional life

SQ is related to individual lives and their relations with life, and the environment. It is related to the mentality and inter-personal relationships. Studies on effects of SQ on personnel health have shown a very positive relationship between spirituality and good health.

The benefits of spirituality or being spiritual are

- Lower diseases prevalence and longer life
- Better response to treatment
- Lower depression rate.

■■■■

Bibliography

Alex, G, Zijp, W and Byerelee, D. 2002. Rural Extension and Advisory Services - New Directions. Rural Development Strategy Background Paper No. 9, Agriculture and Rural Development Department, The World Bank, Washington, DC.

APO. 2006. Enhancement of Extension System in Agriculture, Sharma, VP (ed.). Report of the APO Seminar on Enhancement of Extension System in Agriculture, Pakistan, 15-20 December.

Bashir, Sajitha, 1994. "Public Versus Private in Primary Education: Comparison of School Effectiveness and Costs in Tamilnadu," PhD Thesis. London School of Economics.

Benor, D. and Harrison, J.Q. 1977. Agricultural Extension - Training and Visit System, World Bank, Washington.

Berdegue JA and German Escobar. 2001. Agricultural Knowledge and Information System and Poverty Reduction, *AKIS Discussion Paper* (Jan 2001). The World Bank, Washington DC.Article available at www.rimisp .org/FCK editor/User Files/File/documentos/docs/pdf/-0115-000824-akisandpovertyrevisedfinal.pdf.

Braun-Munzinger, Corina. 2005. "Education Vouchers - an International Comparison," Center for Civil Society Working Paper.

Dahama, O.P. and Bhatnagar O.P. 1998. Education and Communication for Development. Oxford and IBH Publishing Co. Pvt. Ltd., New Delhi.

Das, P. and Hansra, B.S. (Eds.) 1999. Krishi Vigyan Kendra - A Reality, Division of Agriculture Extension, ICAR, New Delhi.

De, Anuradha; Noronha, Claire and Samson, Meera. 2001. "India: Private Schools and Universal Elementary Education," South Asia Education Sector, Technical Working Paper No. 3. World Bank.

Department of Rural Development 1988. Integrated Rural Development Programme and Allied Programme: A Manual. Ministry of Agriculture, Govt. of India, New Delhi.

Ensminger D. 1957. A Guide to Community Development. Ministry of Community Development and Cooperation, Govt. of India, New Delhi.

Farrington J, Christoplus I, Kidd AD and Beckman M. 2002. Extension, Poverty and Vulnerability: The Scope for Policy Reform (Final Report of a study for the Neuchatel Initiative), Working Paper, 155, Overseas Development Institute, UK.

Horton P.B. and Hunt C.L. 1976. Sociology. McGraw-Hill Kogakusha Ltd., Tokyo.

Jalihal, K.A. and Veerabhadraiah, V. 2007. Fundamentals of Extension education and management in extension. Concept Publishing Company, New Delhi.

Kattar Singh. 1999. Rural Development – Principles, Policies and Management. Sage Publications India Pvt. Ltd., New Delhi.

Kelsey L.D. and Hearne C.C. 1967. Cooperative Extension Work. Cornell University Press, Ithaca, New York.

Kelsey, L.D. and C.D. Hearne. 1963. Cooperative Extension Work, Cornell University Press, New York.

Koontz H. and Weihrich H. 1988, 2005. Management. Tata McGraw-Hill Publishing Company Limited, New Delhi.

Leagans J.P. 1961. Characteristics of Teaching and Learning in Extension Education, Extension Teaching Methods, The Communication Process and Programme Planning to Meet People's Need, in Extension Education in Community Development. Directorate of Extension, Government of India, New Delhi.

National Commission on Agriculture 1976. Report, Part I, Review and Progress and Part XI, Research Education and Extension. Ministry of Agriculture and Irrigation, Government of India, New Delhi.

National Institute of Agricultural ·Extension Management.(1999). Agricultural Technology Management Agency (MANAGE-NATP Series-2). Hyderabad: MANAGE.

Neuchatel Group. 2002. Common framework on Financing Agricultural and Rural Extension, Neuchatel Group, Swiss Centre for Agricultural Extension and Rural Development, Lindau.

Prasad, C., Choudhary, B.N. and Nayar, B.B. 1987. First-Line Transfer of Technology Projects. Indian Council of Agricultural Research, New Delhi.

PROBE Team. 1999. Public Report on Basic Education in India. Oxford University Press, New Delhi.

Ray, G.L. 2012. Extension Communication and Management, Kalyani Publishers, New Delhi - 110002.

Reddy, Adivi. A. 2001. Extension Education. Sree Lakshmi Press, Bapatla, A.P.

Robert. M. Ritchie. 2005. Carving the niche. Volume 25 number 3 available at http://www.joe.org/joe/2005 fall/a3.html.

Rogers, F. Halsey 2006. "Missing in Action: Teacher and Health Worker Absence in Developing Countries." Journal of Economic Perspectives, 20(1), pp. 91-116.

Saravanan R. 2001. Privatization of Agricultural Extension. MANAGE, Hyderabad.

Extension: A Strategy of Social Change for Next Millennium. Paper presented at National Symposium on Social Transformation in Rural Sector, Feb. 1-3. Palli Siksha Bhavana, Visva-Bharathi, Shriniketan

Singh R. 1998. Management of Extension Education: New Dimensions of Extension Strategy in Changing Agricultural Scenario. Indian Journal of Extension Education, Vol. 34, No. 1 & 2.

Singh, A.K. 2000. Agricultural Extension. Impact and Assessment. Agribios (India), New Delhi.

Sulaiman, V.R. and Hall, A.J. 2004. Towards Extension plus: Opportunities and Challenges for Reform. NCAP Policy Brief No. 17, National Centre for Agricultural Economics and Policy Research, New Delhi, India. 4 pp.

Van den Ban A.W. and Hawkins H.S. 1988, 1996. Agricultural Extension. Longman Scientific and Technical, England: and Blackwell Science Ltd., London.

Wang G. and Dissanayake W. 1984. Indigenous Communication Systems and Development: A Reappraisal in Continuity and Change in Communication Systems, (eds.), New Jersey

Woessmann, Ludger. 2006. "Public-Private Partnerships in Schooling: Cross-Country Evidence on their Effectiveness in Providing Cognitive Skills, "Mobilizing the Private Sector for Public Education (Conference Volume- forthcoming). Kennedy School of Government, Harvard University.

World Bank. 1996. Development in Practice: Improving Women's Health in India. Washington, D.C.